Carl Bly

St. Paul, MARCH, 1993

Land from the Sea

Land
from the
Sea *The Geologic*
Story of South Florida

John Edward Hoffmeister

UNIVERSITY OF MIAMI PRESS
Coral Gables, Florida

Fifth Printing 1988

Designed by Bernard Lipsky

Manufactured in the United States of America

Library of Congress Cataloging in Publication Data

Hoffmeister, John Edward, 1899-
    Land from the sea.

    Bibliography: p.
    1. Geology   Florida.   I. Title.
QE100.S68H63   557.59   73-20120
ISBN 0-87024-268-7

Illustration credits: Dr. Mahlon Ball: figures 6 & 23. Cooke, 1939:
figure 3. Dillon Aerial Photography, Inc.: figure 50. Dr. Robert Dun-
ham: figure 7. Ginsburg, 1956: figure 64. James Hitch: figures 40 & 65.
Richard Marra: figures 21, 24, 25, 27, 30, 35, 39, 41, 54, 55, 69, 70,
71, & 72. Dr. Donald Marszalek: figures 62 & 63. Dr. Norman L.
McIver, Shell Development Co.: figures 66, 67, & 68. Miami-Metro
News Bureau: figure 9. Ocean Industries, Inc.: figure 22. Puri and Vernon, 1964: figure 1. Schroeder, 1958: figure 74. E. A. Shinn: figures
36, 37, & 44. Smith, 1971: figures 38 & 42. Stockman, Ginsburg, and
Shinn, 1967: figure 61. U.S. Coastal and Geodetic Survey: figure 19.
Wanless, 1969: figures 56 & 57.

To Ruth

Contents

Figures *9*

Preface *13*

Acknowledgments *15*

1 South Florida's Geologic Formations *19*

2 The Florida Mainland *27*

3 The Sedimentary Barrier Islands *50*

4 Corals and Coral Reefs *63*

5 The Florida Reef Tract *70*

6 The Florida Keys *83*

7 Bays of the East Coast *97*

8 Florida Bay *106*

9 Florida's West Coast *116*

10 Water Resources of South Florida *122*

11 A Few Things to Remember *136*

Selected Bibliography *139*

Index *141*

Figures

1. Floridan Plateau *20*
2. Geologic map of South Florida *22*
3. Shoreline of the Pamlico Sea *25*
4. Physiographic map of South Florida *28*
5. Silver Bluff exposure *30*
6. Loose ooids *31*
7. Enlargement of ooids *31*
8. Oolitic rock *32*
9. East facade of Vizcaya showing oolitic rock *33*
10. Spoil bank in the Everglades *34*
11. Multilaminate bryozoan colony *35*
12. Cellular structure of bryozoan *36*
13. Enlargement of knob section *36*
14. Bryozoan colony encrusting seaweeds *37*
15. Exposed bryozoans in the Everglades *38*
16. Cross section of Miami limestone *39*
17. Physiographic features of southern Florida *40*
18. Mirror image relationship of southern Florida and the Great Bahama Bank *42*
19. Unstable oolite ridge, Great Bahama Bank *43*
20. Comparisons of fossil and Recent bryozoans *44*
21. Transverse valleys in the Atlantic coastal ridge *46*
22. Oolite being mined at Ocean Key *48*
23. Internal structure of ooids *48*
24. The sedimentary barrier islands *51*

25. Bear Cut area   *54*
26. View of rock reef at low tide   *54*
27. Diagram of reef platform   *55*
28. Lattice structure of rods   *56*
29. Lateral and vertical roots of black mangrove   *57*
30. Root structure of black mangrove   *58*
31. Exposed black mangrove roots   *59*
32. Platform and rods exposed by erosion   *59*
33. Cross section of pneumatophore   *61*
34. Cross section of fossil pneumatophore   *61*
35. Coral polyp   *64*
36. Living staghorn coral   *65*
37. Brain coral colony   *66*
38. Distribution of major coral reefs   *68*
39. The Florida reef tract   *70*
40. Hydrographic chart of the reef tract   *71*
41. Cross section of the reef tract   *72*
42. Star coral, enlarged   *74*
43. Outer reef corals   *75*
44. Star corals with multilobular heads   *76*
45. Corals of the patch reefs   *77*
46. Finger corals of the back reefs   *79*
47. Coral growth rate experiment   *80*
48. Measuring star corals   *81*
49. Measuring staghorn corals   *81*
50. Aerial view of the Florida Keys   *84*
51. Geologic map of the Florida Keys   *85*
52. Exposure of Key Largo limestone at Windley Key   *87*
53. Section of Key Largo limestone at Cross Key Canal   *90*
54. Diagram of coral reef tract 100,000 years ago   *94*
55. Diagram of present-day coral reef tract   *95*
56. Bathymetric map of Biscayne Bay area   *98*
57. Bedrock topography of Biscayne Bay area   *100*
58. Quartz grains, angular and clear   *102*
59. Quartz grains, rounded and cloudy   *103*
60. Mudbanks in Florida Bay   *107*

61. Seaweed growth in Florida Bay *110*
62. Morphology of *Penicillus*, a common green algae *111*
63. Microscopic crystals of aragonite *111*
64. Florida Bay and reef tract *114*
65. Florida's western coast *117*
66. Ten Thousand Islands area *118*
67. Oyster beds at low tide *119*
68. Mangroves encroaching on oyster beds *120*
69. Diagram of water table in wet and dry seasons *123*
70. Geographic extent of the Biscayne aquifer *124*
71. Cross section of the Biscayne aquifer *126*
72. Cross section showing aquiclude and artesian wells *127*
73. South Florida's canal and levee system *130*
74. Progression of saltwater encroachment, 1904-1953 *133*
75. Saltwater intrusion in 1967 *134*

Preface

This earth on which we walk, plant our crops, build our homes and roads, and dig into for swimming pools, boat slips, and canals is truly the good earth, providing us with the essentials and luxuries of life.

Man works this earth for all he can, showing great interest when oil or other important mineral deposits are found, but in general remaining indifferent to the earth itself. After all, it's only made of rocks.

It is a fact, however, that there is a storehouse of interesting things in every part of our earth if we would only take a little time to examine and think about it. The sands of the deserts, the rocks of the Rocky Mountains, the mud on the bottom of Florida Bay, and the stratified rock formations of the Great Plains are filled with mysteries, some of which have been solved and some of which still wait for their solutions.

Astronauts have gone to the moon at tremendous expense and great personal danger. They have returned with rocks as their chief trophies. Most of the rocks brought from the moon are in scientific laboratories, the objects of concentrated study. A few moon rocks have been placed under glass in well-guarded rooms, and thousands of persons have gazed on them in awe.

This interest in moon rocks is amazing when compared with the indifference and even contempt with which man regards the rocks of his own earth. This indifference could be understood if men were more familiar with their local rocks, but very few people know anything about the rocks of the earth, and fewer still have even looked at them with care.

Have you ever taken the trouble to go into your yard to pick up a piece of the rock on which your home is built and look at it carefully? If not, let me suggest that you do so. You may not see much at first. The particles of the rock you look at may be too small for you to see without some magnification. But with only a magnifying glass or a small hand lens, the chances are that you will be amazed at what you see in just an "ordinary rock." With a few books and the simplest equipment, and preferably someone to guide you initially, your own earth may turn into a whole new world for you.

Acknowledgments

This book is based upon field and laboratory research carried out by the author and his associates in Southern Florida over a period of about ten years. The National Science Foundation has provided funds for the support of much of the research presented here. The results of the work have already been published in scientific journals. Since the area is geographically unique and of unusual geologic interest it seemed desirable to write the story in words that could be understood by the general public.

It has been my privilege to have the Rosenstiel School of Marine and Atmospheric Science of the University of Miami as my headquarters during my years of work in Florida. Dr. F. G. Walton Smith, dean emeritus of the school, and his wife, Mrs. May Smith, executive secretary of the International Occanographic Foundation, have extended their gracious friendship to me throughout this time.

Many persons have contributed substantially to the project and I wish to acknowledge their help. First, I wish to thank my valued friend and co-worker, Dr. H. Gray Multer, professor of geology at Fairleigh-Dickinson University. His help in all aspects of the undertaking has been indispensable. In addition, I am grateful to a number of others who have been outstanding in providing help, especially in the exploratory stages of the work. Dr. Robert N. Ginsburg of the Rosenstiel School first introduced me to the coral reef tract and has been helpful in a thousand ways over the past ten years. Mr. E. A. Shinn of the Shell Oil Company, coral authority and expert diver, has been my right-hand man in coral reef work. Mr. Axel Olsson and I jointly discovered the importance of bryozoans in the Miami limestone. Mr. Kenneth Stockman, my research assistant, was most helpful, as were Dr. John Milliman of the Woods Hole Oceanographic Institution, Dr. Alan Cheetham and Dr. Harry S. Ladd of the U.S. National Museum, Dr. Taylor Alexander and Mrs. Mabel Miller of the University of Miami Department of Botany, Dr. Norman Newell of the American Museum of Natural History in New York, Dr. Harbans S. Puri of the Florida Geological Survey, and Dr. John Wells of Cornell University.

Major contributions have been made by Mr. Fred Meyer of the U.S. Geological Survey in the chapter on water resources and by Dr. Harold Wanless, of the Rosenstiel School, for his help on the chapter concerning the bays of the east coast.

To my many associates and colleagues at the Rosenstiel School I give thanks for help and encouragement: Dr. Cesare Emiliani, chairman of the Department of Marine Geology; Dr. Mahlon Ball; Dr. Frederick Bayer; Dr. Wayne Bock; Dr. Stefan Gartner; Mr. Donald Heuer; Mr. James Hitch; Dr. Oiva Joensuu; the late Dr. Fritz Koczy; Mr. Richard Marra; Dr. Donald Marszalek; Dr. Donald Moore and his gracious wife, Cynthia; Dr. Frederick Nagle; Dr. Conrad Neumann; Mr. Robert Stearns; Dr. Anitra Thorhaug; Mrs. Ruth Trencher; Mrs. Faith Turnbull; Mr. Robert Work; and Mr. Bernard Yokel. I also thank Dr. Harris Stewart, director of NOAA's Atlantic Oceanographic and Meteorological Laboratories.

In addition to these I thank Mrs. Jean Bradfisch of the International Oceanographic Foundation for her help in arranging and editing much of the manuscript, Mrs. Hoppes Barker and Mr. Erik Speyer for their art suggestions, and Mrs. Bette R. Kendrick for secretarial help. It is with warm gratitude that I acknowledge the kindness of my friend and former student Mr. Robert Tucker, president of the Rochester Polychrome Press, who has contributed the jacket for this book.

These and many more have been my cheering section.

Land from the Sea

1| South Florida's Geologic Formations

There are two major steps in the geologic study of the earth. The first is to look at its rocks and find out what they are made of. The second step is to try to understand the conditions under which these rocks were formed and how and when they became what they now are. Like most scientific work, geology involves detective work. Like detectives, geologists first try to get the facts, and then, as they pick up various clues, they finally reach a solution.

Peninsular Florida is shaped like a long narrow finger pointing to Cuba and separating the waters of the Gulf of Mexico from the Atlantic Ocean. This extension of the mainland appears to be a rather fragile divider of these two huge bodies of water, but appearances from the surface are deceiving. What we see as the state of Florida is only the upper part of a much more substantial barrier. If the sea level were lowered about 50 fathoms (300 feet), the width of the finger would be greatly enlarged and it could be seen that Florida rests on a broad, flat platform—the Floridan Plateau (fig. 1). The 50-fathom line marks the edge of the relatively shallow continental shelf. Beyond this line at the edge of the plateau the sea bottom drops to much greater depths.

There is reason to believe that the Floridan Plateau has been in existence without notable change for many years. Compared to many other parts of the earth, it has remained a rather stable unit. Its rocks have experienced relatively little bending or fracturing. The result is that earthquakes are practically unknown.

Fig. 1. *The Floridan Plateau with generalized locations of landforms.*

The changes that have occurred in the Floridan Plateau have been largely superficial, but at the same time they are most important. It is known, for example, that the level of the sea has fluctuated up and down all over the world and is still doing so. When the sea level rose, the land exposed above water became correspondingly smaller; at times Florida disappeared entirely. When the sea level dropped, more and more of the plateau became land, and in the not too distant past Florida was as broad as the present surface of the plateau. We can therefore speak of the exposed Florida and the hidden Florida.

When changes of this type occur, they are inevitably accompanied by others. When the land is high, erosion is active and new landforms, such as river valleys, are made. When the sea extends over the land, new rock deposits, chiefly limestone in Florida, are formed on the shallow sea bottom, largely from the accumulation of shells and other skeletal materials of the organisms that lived in the sea. Various places in Florida are covered by such deposits.

Florida is special in many ways, especially for a person who likes the sea. Next to Alaska, it has the longest coastline (1,350 miles) of any of the states. Its surrounding waters teem with marine animals and plants that produce large amounts of limestone. It has a flourishing living coral reef tract and a unique chain of islands, the Florida Keys, which arc made chiefly of an ancient coral reef. These features and many more make it outstanding.

The southern tip of Florida is a warm, semitropical area attractive to many people, both residents and visitors. But long before man appeared here, many heat-loving animals and plants occupied the land and the sea around it. Among the most important of these were the coral animals that thrive only in tropical and subtropical climates.

Today corals occupy the shallow water bordering the southeastern coast of Florida from about Miami to Key West. This is the living coral reef tract. At times in the past, the warm coral-bearing waters extended over what is now land and left in some places patches of coral reefs. This southern coral-bearing area of living and fossil coral with its associated animals and plants might be called the coral reef zone of Florida. It makes a natural unit that is different in many respects from the rest of the state. This coral reef zone, a major concern of this book, is the southern part of the area that the Florida Geological Survey has designated as the Southern Zone of the Coastal Lowlands (fig. 1).

Geologic History of the Formations

The rock formations of southern Florida are among the youngest in the country. Our best estimate at the present time is that the crust of the earth originated about 3,000 million years ago. Thus the rocks formed at that time are the oldest known to man. In contrast, the oldest rock formation shown on the geologic map (fig. 2) and in the Geologic Time Table (p. 23) is the Tamiami limestone, which was formed here only about 6 million years ago. It belongs to the Miocene epoch of geologic time and is the rock that can be seen

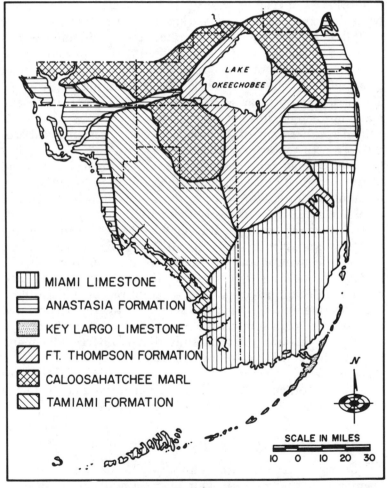

Fig. 2. *Geologic map of South Florida showing location and surface area of each formation.*

GEOLOGIC TIME TABLE

Geologic Age (Epoch)	Formation	Characteristics	Thickness (in ft.)	Age (in yrs.)
	Pamlico Sand	Quartz sand, white to black or red. Mantles part of Miami and Anastasia formations. Occurs in sand dunes and old beach ridges.	0-60	100,000
	Miami Limestone	White to yellowish. Massive to stratified and cross-bedded. Oolitic and bryozoan facies.	0-40	100,000
	Anastasia Formation	Coquina, sand, calcareous sandstone, and shell marl. Probably composed of deposits equivalent in age to marine members of Fort Thompson and Miami limestone.	0-120	100,000+
Pleistocene (Ice Age)	Key Largo Limestone	Coralline reef rock. Hard and cavernous. Interfingers with bryozoan facies of Miami limestone and probably with Fort Thompson. Excellent aquifer.	0-200+	100,000+
	Fort Thompson Formation	Alternating marine and freshwater marls, limestones, and sandstones. Main component of Biscayne aquifer in eastern part of Dade and Broward counties. Northern extension much less permeable.	1-150	Upper Part 100,000+
Pliocene	Caloosahatchee Marl	Sandy marl, clay, silt, sand, and shell beds. Poor to fair aquifer.	0-25	2,000,000+

Miocene	Tamiami Limestone	Cream, white and greenish-gray clayey marl, silt, and shelly sands and sand marl, locally hardened to limestone. The upper part, where permeability is high, forms the basal part of Biscayne aquifer. The lower and major part of the formation is of low permeability.	0-100	6,000,000

especially well in certain areas of western Dade and Collier counties. Next in age is the Caloosahatchee marl of the Pliocene epoch that outcrops in the northern part of the area and is somewhat older than two million years.

All the other formations belong to the Pleistocene, or Ice Age, which began about two million years ago and lasted right up to what is known as the Recent epoch, which began about 11,000 years ago and includes the present time. Although the Fort Thompson, the Key Largo, the Anastasia, and the Miami formations are all of about the same age, they were deposited under different environmental conditions. This accounts for the difference in their composition. For example, the upper part of the Key Largo limestone, which is confined chiefly to the Florida Keys, was forming at the same time the Miami limestone was developing on the mainland. This can be seen by the fact that the Key Largo limestone interfingers with the Miami. In all probability the lower part of the Key Largo limestone is also contemporaneous with the Fort Thompson formation.

The Anastasia formation is found bordering both the Atlantic and Gulf coasts in the northern part of South Florida. Near the border between Broward and Palm Beach counties the Anastasia and the Miami grade into each other and the Anastasia replaces the Miami as the backbone of the Atlantic coastal ridge (see fig. 2). On the Gulf coast the Anastasia is exposed as a narrow strip from a point north of the Ten Thousand Islands in Collier County to about 10 miles north of Sarasota (Puri and Vernon, 1964, p. 282).

Figure 3 shows the shape of Florida about 100,000 years ago when the sea stood about 25 feet above its present level. At this time the coral reefs of the Keys and the oolite mound that later was to become the Miami limestone were being formed. The waves of the sea beating against the coasts of a much smaller Florida formed, by erosion and deposition, a broad terrace at that level—the Pamlico terrace. The Pamlico is a grayish white to dark brown quartz sand that contains some carbon. In places it becomes a hard sandstone. It also may contain many fossils. When the sea level lowered, due to the start of the Wisconsin Ice Age, the Pamlico sand was left behind. It was scattered irregularly by wind and rain over the older formations of southern Florida. In most places where it occurs it is only a foot or two in thickness, but locally it may be heaped into beach ridges and dune deposits of greater

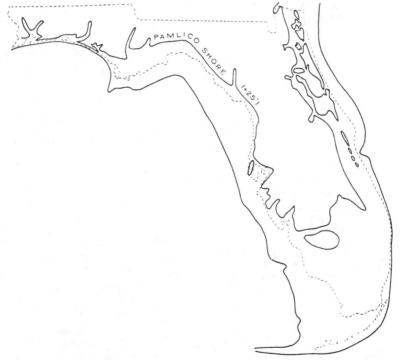

Fig. 3. *Shoreline of the Pamlico Sea when the sea level was 25 feet higher than at present.*

thickness. This scattered pattern makes it difficult to show on a geologic map. The Pamlico sand is the youngest Pleistocene formation in the region.

The geologic map and the brief descriptions of the rock formations that appear at the surface provide a background for understanding the origin of the important geographic sections of South Florida. Beginning with the mainland, the major geologic events of each section will be described and developed in chronological order.

2| *The Florida Mainland*

The topography of the mainland of the southern tip of Florida is uncomplicated. No part of it is very far from the coast, and much of it is about as low as a place can be and still be called land. It appears to possess about the same low, flat, featureless surface throughout its extent. No hills stand up above a rather monotonous level, but here in the coastal lowlands, it is surprising to see what a difference is made by only a few feet in elevation.

This region is naturally divided into two main topographic areas: (1) the Atlantic coastal ridge, and (2) the Everglades, including the Big Cypress Swamp (fig. 4). The Atlantic coastal ridge borders the Atlantic shore as a narrow ridge with a smooth to undulating surface from the Georgia state boundary to Homestead. Upon it are located the main cities of eastern Florida, including Fort Lauderdale and Miami. The highest point on the ridge is about 20 feet above sea level in the Coconut Grove section of Miami (fig. 17). This is the oldest part of Miami, and there is no doubt that the early settlers chose this area primarily for its elevation—a place where they were relatively safe from sea action and mosquitos. A few vertical feet make a great deal of difference during a hurricane when the rising sea level floods lower areas. From the ridge the land slopes eastward to the nearby Atlantic and westward, more gently, to the Everglades. The low, broad expanse of the Everglades is about 10 feet above sea level in southern Broward County. To the south it drops impercep-

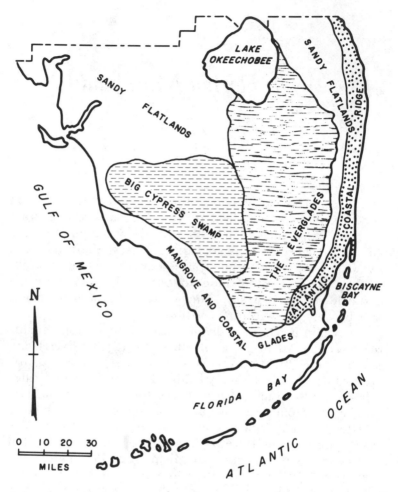

Fig. 4. *Generalized physiographic map of South Florida.*

tibly to a five-foot level in Dade and Monroe counties. As the southern and western coasts are approached, the land slowly drops to sea level and a jagged, reticulate shoreline with innumerable mangrove islands and small, tortuous creeks results. The Atlantic coastal ridge has been called the rim of the Everglades.

The Rocks of the Atlantic Coastal Ridge

In many respects the Atlantic coastal ridge is one of the most important parts of South Florida. The majority of the population is concentrated on this narrow eastern ridge. If the persons who live here were asked what kind of rock is beneath their feet, the majority would immediately say "coral rock." If they saw a stone wall or building of the same material many would undoubtedly proclaim it to be "coral." It would surprise them to learn that there are no corals or coral reef rocks in the local surface formations. If one would dig beneath the surface—down to several feet beneath sea level—he might uncover a few corals here and there, but we actually would have to travel a few miles away from the mainland to locate this valued material.

To make a correct identification our best information comes from the so-called bedrock, the hard, solid rock that in most places is covered by loose material or soil of varying thicknesses. In some regions the covering material is so thick that the bedrock is not exposed, but this is far from the case in southern Florida. This is a region of thin soils, so thin that it is difficult to plant shrubs and small trees without the laborious job of digging into hard rock. One small advantage of this condition is that heavy cars may be driven directly over a beautiful lawn with little fear of damaging it. In thicker soil deep ruts would result.

Outcrops or exposures of bedrock occur often along the coastal ridge, many at excavations for the foundations of buildings, roads, swimming pools, and so on. Wherever there is an excavation, no matter how shallow, it is practically certain that the exposure will reveal bedrock. Canals that have been cut through the coastal ridge permit water from the interior to flow eastward and empty into Biscayne Bay. The vertical sides of these canals offer excellent views of bedrock. One of the best viewing sites is along the sides of the Coral Gables Canal, which passes by Coco Plum Circle at the southern end of LeJeune Road. The exposure that has attracted the most attention, however, is that of the Silver Bluff section along Bayshore Drive in Coconut Grove (fig. 5).

Fig. 5. *Cross-bedded oolite of Miami limestone at Silver Bluff, South Bayshore Drive, Coconut Grove, Miami. The wave-cut notch is about 8 feet above present sea level.*

A great deal of information about the origin and past history of an area can be obtained by simply looking at a good exposure of bedrock and examining its contents, even if this is done in a rather superficial way. The Silver Bluff exposure is a good one to observe. Here we first see that the rocks are distinctly layered. Next, the layers are lying at different angles to each other—some dipping in one direction, some in another, some almost horizontal. This is known as "cross-bedding." A closer examination with the aid of a lens, even a small hand lens, shows that the layers are composed of billions of tiny, spherical sand grains that are held together by a kind of cement. A few drops of a weak solution of an acid, such as hydrochloric acid, will cause the rock to effervesce.

These clues tell us that the layered rock was formed under water and that it was originally made of loose, unconsolidated sand grains. The grains, at first suspended in the water, later filtered down to the bottom where they formed a layer. As more grains appeared, they in turn were deposited on those layers previously formed. In this way, layer after layer was deposited. If the body of water in which the layers of sediment were building up had been a quiet one, the layers would have continued to be deposited in an orderly, horizontal fashion. The water was turbulent, however, with currents running in different directions; some of the layers were tilted at an angle to the bottom and others were deposited at different angles, and a cross-bedded condition developed.

Viewed through a hand lens the sand grains of this formation are seen to be shaped like tiny spheres. The fact that these spheres effervesce or give off a gas (carbon dioxide) when acid is applied indicates that they are made of limestone (calcium carbonate), i.e., are calcareous. Calcareous spheres of this type are known as "ooids" because of their resemblance to small eggs, such as those laid by fishes (figs. 6 and 7), and a rock that is made of them is known as an "oolite," or is called oolitic (fig. 8).

These ooids were loose and unconsolidated as they were

Fig. 6. *Loose ooids from the Great Bahama Bank.*

Fig. 7. *Ooids greatly enlarged. Egglike shapes give ooids their name.*

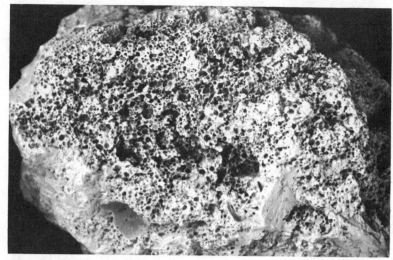

Fig. 8. *Oolitic rock, Miami limestone. Most of the ooids at the surface have been dissolved away, but the calcareous cement remains.*

deposited layer after layer on the sea bottom. But at a much later time when the sea level lowered, the layers of ooids became dry land. Then rainwater trickled down between the grains, dissolved some of the lime, and reprecipitated it around the grains as a cement, binding them to each other and forming the hard, solid rock that we see today. In summary, the early environment of this region caused the formation of calcareous ooids that were deposited in layers in a cross-bedded manner until an elongated, north-south submarine mound was formed that eventually became the oolitic rocks of the Atlantic coastal ridge.

Oolite is really a beautiful rock. It weathers in many intricate, mysterious shapes, and it makes a satisfactory building stone. One of the showplaces of Miami, Vizcaya, a magnificent Mediterranean villa located in the southern part of Miami near Coconut Grove, is largely constructed of oolite.

By far the greater part of oolite that is quarried is then crushed and converted into concrete building blocks or used in the making of roads, ready-mixed concrete, and cement. The high-rise apartment and office buildings that fashion the skyline of the cities of the southeastern Florida coast are built largely of this material. So also are thousands of smaller residences—houses built of concrete blocks. All of this building stone is native to the region, and the greater part comes from Dade and Broward counties.

There is an old saying among geologists, "I'd rather own a limestone quarry than a gold mine." The mining of oolite is big business, and quarries are numerous and scattered. One of the advantages of limestone mining here is the relative ease of extraction. It is all open pit mining, and there is practically no overburden to be removed as the rock outcrops at the surface. Because of its homogeneous structure and geologic youth, oolite is a relatively soft rock, easy to crush. In addition, a large percentage of it is mined within a very short distance of the places where it is used, and thus transportation time and costs are minimal. Large quantities of Florida oolite are now being shipped to more distant places.

Quarries, instead of being eyesores on the landscape, are often converted into attractive lakes that become the centers

Fig. 9. *Vizcaya (east facade shown here) is built largely of oolitic rock.*

of residential developments. We can be grateful that oolite, like coral, is very porous. Rainwater seeps quickly beneath the surface and builds up the groundwater supply where it is stored for our everyday use. If the surface rock were more solid and impervious, rain would run off at the surface, gather in streams, and quickly empty into the ocean and be lost.

The Rocks of the Everglades

The coastal ridge slopes gently westward to the low-lying Everglades. A region with an elevation of plus-five feet above sea level is not a promising one in which to examine bedrock. This is especially true when it is covered, as in the Everglades, with mud, water, and extensive growths of sedge. But even here, particularly in times of drought, there are patches of exposed bedrock.

Fortunately, the geologist does not have to rely only on naturally exposed bedrock in his search for the characteristic rock that underlies the Everglades. Numerous canals have been dug throughout the area, and the spoil banks that result reveal clearly the type of rock present (fig. 10). All one needs to do is pick up some broken rock fragments and examine them with a hand lens.

It might be expected that the Everglades rocks would be of

Fig. 10. *Canal and spoil bank in the Everglades.*

oolitic limestone similar to that of the Atlantic ridge. Although there are many ooids present, the Everglades rocks are also found to contain large numbers of massive colonies of what are known as bryozoans.

A bryozoan (which means a "moss animal") is a minute, invertebrate, marine animal, usually less than one millimeter in length, which secretes a small, calcareous cell in which to live. Bryozoans grow together to form calcareous colonies, many of which become a foot or more in diameter and develop a typically knobby structure (figs. 11 and 12). In some respects they are similar to corals, although they are slightly higher in the scale of life, and their skeletons are usually not as attractive as those of corals.

The first animal to start a colony is produced sexually, and from this one new individuals are made by nonsexual budding. Most bryozoans are encrusting animals that attach themselves to other animals and plants. Those that are found in the Everglades apparently encrusted various types of sea grass that grew on the ocean floor. When these grass roots and blades were completely covered by a layer of the bud-

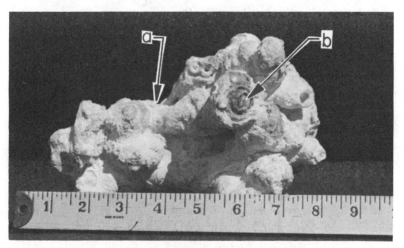

Fig. 11. *Typical knobby specimen of multilaminate bryozoan colony* (Schizoporella floridana) *from the bryozoan facies of the Miami limestone.*

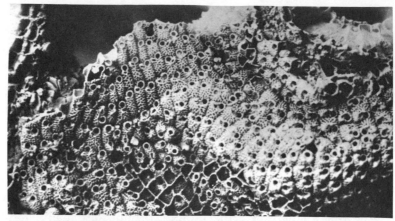

Fig. 12. *Enlarged view of well-preserved cellular structure at point of arrow marked* a *in fig. 11.*

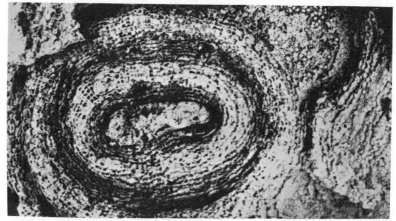

Fig. 13. *Enlarged view of transverse section of knob at point of arrow marked* b *in fig. 11.*

ding, calcareous cells, each of which enclosed a living, soft-bodied animal, new layers encircled the first layer until a fairly thick, tubelike structure was formed. It is not uncommon to find specimens with a dozen or more layers. The term "multilaminate bryozoans" has been applied to those with this structure (fig. 13).

Bryozoan colonies vary tremendously in shape, joining together to form irregular masses. It is evident that the shape depends on the character and form of the encrusted organisms. In the shallow waters around the shores of the Keys and in places in Florida Bay, excellent examples of bryozoan-encrusted sea weeds may be found. Figure 14 shows how this occurs. It is believed that the bryozoan colonies found as fossils in the Everglades rocks were produced in this manner. In practically all specimens of fossil colonies, the knobs and more elongated branches have hollow centers. The original encrusted woody material of the sea plants disappears by decomposition, leaving tubes of multilaminate bryozoan skeletons.

Bryozoans are by far the dominant constituent of the Everglades rocks. It is estimated that in many places they occupy at least 70 percent of the rock by volume (fig. 15). One rather amazing thing is that they all belong to a single

Fig. 14. *Recent specimen of* Schizoporella floridana *from Florida Bay shows how bryozoan colony encrusts seaweed. Arrow points to branch of exposed seaweed.*

Fig. 15. *Spoil bank along a canal in the Everglades. At least 70 percent of the rock specimens are fragments of bryozoan colonies.*

species, *Schizoporella floridana.* In addition to bryozoans, calcareous worm tubes are numerous and comprise in places as much as 10 to 20 percent of the rock by volume. Marine worms of different types obtain limestone from the water and secrete it around their bodies to build tubes in which they live and are protected from other animals. Other important constituents are ooids, like those of the coastal ridge, and also what are known as pellets. The latter are tiny, elliptical, calcareous grains similar in shape to a football. They are believed to have been excreted by marine worms.

Geologic Relationships

The best way to find out what has happened beneath the surface is by core drilling in strategic places. In this process, long metal cylinders about two inches in diameter are drilled into the ground; when returned to the surface they contain a core, or sample of material at different levels below.

Core drilling usually begins at a high point in order to obtain the thickest section for study. A sample core that was taken on the coastal ridge at an elevation of 20 feet showed that the upper 30 feet is composed of oolite, followed by about 10 feet of massive, tubular bryozoans in an oolitic and pelletoidal matrix. Or, in other words, the oolitic rock of the ridge overlies the bryozoan rocks of the Everglades. These two rock units have a great deal in common—they are both limestones and both possess ooids and pellets. Together they have been called the Miami limestone. The upper unit is called the oolitic facies and the lower the bryozoan facies.

Information from extensive field work, including drilling, provides a profile of the two units beneath the surface. Figure 16 shows a generalized east-west cross section from the Atlantic Ocean to the Gulf of Mexico. This cross section shows clearly that the oolitic facies has a thickness under the Atlantic coastal ridge of about 30 feet and that the bryozoan facies at this point is about 10 feet thick. The oolitic facies thins in a westerly direction toward the Everglades as it encroaches over the bryozoan facies, until it disappears at the place where the bryozoan facies reaches the surface. The latter gradually becomes thinner until it disappears entirely near the Gulf.

One interesting feature of the cross section is the appearance in the eastern end of the bryozoan facies of two sheets of coral reef limestone. This is the Key Largo limestone, which apparently extended from the Florida Keys area and encroached on the mainland at some time in the past.

Figure 17 shows the approximate area occupied at the surface by the two facies of the Miami limestone. A large part of the bryozoan facies is covered by a varying thickness of

Fig. 16. *Generalized east-west cross section of Miami limestone showing the relationship between the oolitic facies and the bryozoan facies. Also shown at the extreme right is the interfingering of the Key Largo limestone with the bryozoan facies.*

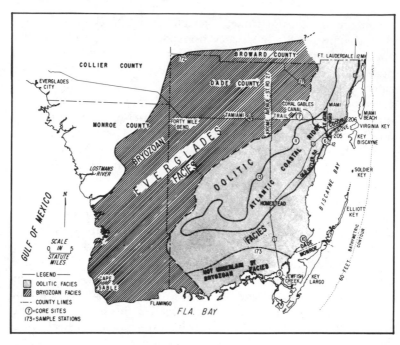

Fig. 17. *Geologic map showing areas covered by the oolitic and the bryozoan facies of the Miami limestone.*

the oolitic facies. The bryozoan facies extends as far north as Fort Lauderdale. West of Fort Lauderdale in the Everglades it has been traced to a point about two miles east of the intersection of Dade and Broward and Collier counties. This is probably near its northern limit. Along the western side it can be traced for over 10 miles west of the Forty Mile Bend of the Tamiami Trail and thence to the southwest, where it reaches the Gulf of Mexico at a point about 10 miles south of Lostman's River. It is not known whether it underlies all of Cape Sable, but Flamingo is one of the most prolific localities.

The size of the area covered by the bryozoan limestone in this region makes it unique. The facies covers an area of at least 2,000 square miles and is certainly one of the most extensive bryozoan limestones in the country.

Present-day Clues to the Past

Knowing that the southern tip of Florida, covered by the Miami limestone of two different types, has an eastern area made of an oolitic facies of inorganic origin and a western area of a bryozoan facies of organic origin gives one the basis for understanding the geological development of this area. To determine how these rocks were formed we need to know the conditions present when the ooids of the Atlantic ridge were forming and when the bryozoans were flourishing in the Everglades. Pertinent questions of study include how deep the water was, what the temperature and salinity of the water were, and how Florida looked during the time these rocks were forming.

Starting with the knowledge that both the organic and the inorganic materials were formed under marine conditions, it is logical to study the ocean for the geologic explanation. Fortunately, we do not have to look very far to find a place in the sea today where similar rocks are being formed—only to the Great Bahama Bank, directly east of Miami on the opposite side of the Straits of Florida (fig. 18).

The western border of the Great Bahama Bank is edged by a north-south line of cays, or islands, including the Biminis and the Cat Cays. East of the cays, parallel to them, and covered by shallow water is a mound or bar of loose oolitic sand (fig. 19). The top of the bar is about six feet beneath sea level. Here individual ooids are being formed in tremendous numbers and are being added constantly to the bar. The bar slopes on its eastern side to the Bank floor, or shelf lagoon, which lies 15 to 20 feet beneath the surface of the sea. Since the water over the bar is very shallow, strong currents running in various directions create a cross-bedded structure in the growing mound. Here, then, is a mound that has all the characteristics of the coastal ridge of the Florida mainland except that the latter is made of hard rock while the Bahama bar is made of loose oolitic sand.

In the Bank area immediately east of the bar in the shallow lagoon the bottom is lined with oolites and pellets, upon which patches of bryozoans flourish in places. These bryo-

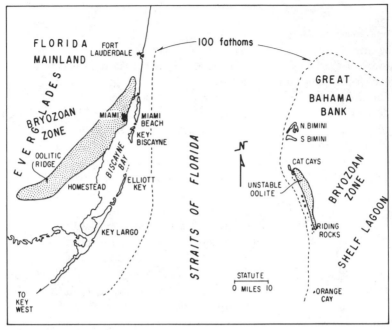

Fig. 18. *Southeastern Florida and the northwestern section of the Great Bahama Bank. This shows the mirror image relationship between the chief topographic and stratigraphic features of southern Florida and their Recent Bahamian counterparts.*

zoans are all of the same species, *Schizoporella floridana,* as those from the Everglades (fig. 20).

The Bahama environment encourages extensive bryozoan growth. The oolite shoal offers protection from large waves from the Straits of Florida, and the strong tidal currents assure a good food supply. The similarities of the two localities are soon evident. One is almost the mirror image of the other. A mirror placed in the middle of the Straits of Florida between Miami and Bimini would reveal similar features in reverse order (see fig. 18).

Starting from the coastline and progressing eastward, the outstanding features in the Bimini area are: (1) the cays; (2) the underwater mound of loose or unstable cross-bedded

Fig. 19. *Nine-lens aerial photograph of the unstable oolite ridge, Great Bahama Bank. The northern part of the ridge, from South Cat Cay at the top of the photograph to Sandy Cay in the south, is a distance of about 9 miles. Transverse tidal channels with deltaic ends show clearly. Bryozoans grow in somewhat deeper water (darker area) east of the ridge. White spots on left are puff clouds.*

Fig. 20. *Comparison of a fossil colony of* Schizoporella floridana *from the Miami limestone with a Recent specimen from the lagoon of the Great Bahama Bank. Top photo shows typical tubular fossil growth form from the bryozoan facies of the Miami limestone. Bottom photo shows similar tubular growth form of a Recent specimen from the Great Bahama Bank.*

oolite; and (3) the bottom or shelf lagoon of the bank where bryozoans flourish in deeper water. On the Florida side from east to west are: (1) the Florida Keys; (2) the Atlantic coastal ridge of cross-bedded oolite; and (3) the low-lying Everglades with its abundance of fossil bryozoans. There is little question that the conditions that are now producing the sediments and living organisms in the Bahamian area are nearly identical with those that existed over the southern tip of Florida about 100,000 years ago.

Our Bahamian counterpart tells us what happened in southern Florida. The level of the ocean in that time must have been high enough to have covered the highest part of what is now our Atlantic coastal ridge. Thus, the sea level at that remote time was about 20 to 25 feet higher than it is today. A mound or bar of loose oolitic sand was built up to nearly sea level, and strong tidal currents created a cross-bedded condition in the mound as it was being formed.

At the same time the mound was forming, bryozoans flourished in the somewhat deeper water over the wide shelf to the west where the Everglades is now located. The presence of these and other marine organisms indicates that the ecologic conditions of the shallow sea were very similar to those of today. Certainly the water temperature was the same and so also was the salinity (about 37 to 39 parts per thousand).

When the level of the ocean was 25 feet higher than it is today, about 100,000 years ago, the mainland of Florida was certainly much smaller and possessed a totally different coastline. The sea remained at this level for a long time, and the old Pamlico shoreline (see fig. 3) can be clearly recognized in many places by the presence of wave-cut bluffs, beaches, ancient bars, marine terraces, and accumulations of seashells.

After many years the ice sheets in the polar areas began to increase in size, taking water from the ocean and lowering the sea level. The oolitic mound and the bryozoan flats slowly emerged to become land. Rainwater percolating downward eventually precipitated calcite in the porous spaces and bound the loose material together to form hard rock. The oolitic mound eventually became the Atlantic coastal ridge, and the bryozoan region to the west became the Everglades.

The Atlantic coastal ridge south of Miami is cut by a series of shallow valleys running parallel to each other and at right angles to the long axis of the ridge (fig. 21). On close examination it is found that the valley floors are flat—only two or three feet lower than the oolitic divides between them. The material deposited in the valleys is made of a limey marl four or five feet thick. This material, called Perrine marl, is well

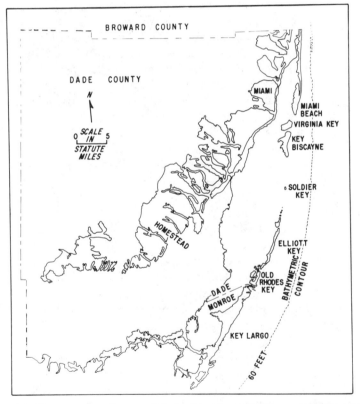

Fig. 21. *Transverse valleys or "glades" in the Atlantic coastal ridge. The valleys are comparable to the tidal channels of the unstable oolite ridge of the Bahama area.*

known by agriculturalists of the region. It is utilized for the production of winter vegetables, especially Irish potatoes.

The geologic explanation of these valleys, locally called "glades," can be obtained by looking at the counterpart of the valleys in the unconsolidated oolite mound of the Bahamas, where submarine tidal currents have cut channels running transverse to the mound (fig. 19). It is believed that similar channels were cut long ago in what is now the Atlantic coastal ridge as it was forming as a mound of loose ooids beneath the sea. When the sea level lowered and the mound was exposed, the loose ooids were cemented together to form the hard, oolitic Miami limestone. The original channels were maintained in the solid rock, and more recently marl was deposited in the ancient valleys.

Oolite Mining

Oceanographers have stressed the importance of the ocean as a source of food and mineral supplies. A good example of the latter is the oolite mining development recently begun in the unconsolidated mound of the Bahama Bank. At a distance of about 20 miles south of Bimini, a new key known as Ocean Key has been formed from oolite dredged from the shallow sea floor. This has become the operational center of the oolite mining industry (fig. 22). From a potential supply of about 20 billion tons, dredged oolite is loaded on ships and widely distributed. The loose ooids are made of almost pure aragonite (a form of calcium carbonate) and are used for many purposes—for cement manufacturing, glass making, agriculture, and even as an element in animal foods.

Origin of the Bahamian Ooids

Microscopic examination shows that an ooid has a small nucleus surrounded by layers of calcium carbonate (fig. 23). The nuclei may be made of tiny fragments of any type of rock that is in suspension in water. It may be made of quartz grains as well as of calcite. The water from the deep Gulf Stream rushing over the shallow Great Bahama Bank experi-

Fig. 22. *Aerial view of oolite mining operation at Ocean Key. Darker color indicates dredged area.*

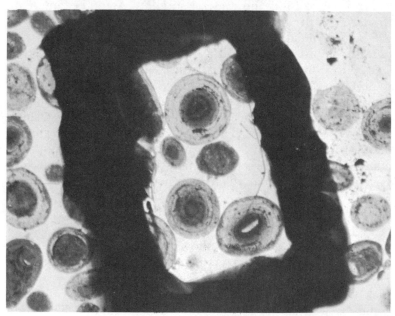

Fig. 23. *Ooids greatly enlarged, showing internal structure.*

ences an increase in temperature and salinity and thus a decrease in calcium carbonate solubility. The rather abrupt change from deep water at the margin of the Bank to the shallow water of the Bank's surface is accompanied by an increase in velocity and turbulence of the current. The current, thus agitated, drives off excess carbon dioxide and further reduces the solubility of the calcium carbonate. The supersaturated water becomes warmer and more agitated and is subjected to evaporation as it moves across the shoals. This causes the precipitation of layers of calcium carbonate around any small fragments in suspension, thus causing the formation of typical ooids.

3| *The Sedimentary Barrier Islands*

The southern tip of Florida is completely surrounded by hundreds and hundreds of islands. The most significant may be grouped as follows: (1) the sedimentary barrier islands; (2) the Florida Keys; and (3) the Ten Thousand Islands.

The sedimentary barrier islands, the most important of which are Miami Beach, Virginia Key, and Key Biscayne, are a series of small, low islands strung out in a north-south direction parallel to the Florida mainland (fig. 24). They are bordered on the east by the Atlantic Ocean and on the west by Biscayne Bay. Key Biscayne is at the southern end of the chain.

The islands were made by a combination of nature and man; nature began the job and man completed it. The islands began as a chain of sandbars. Loose sand particles were carried by ocean currents from the north and dropped in a shallow-water zone near the mainland. An elongated north-south sandbar was built up toward sea level. Ocean waves and wind action caused them to be mounded above the sea and turned into islands.

It did not take long for the low, sand islands to be covered by shrubs and trees, the most important of which are mangroves. When men first decided to develop the area, in 1912, they had to deal with a mature mangrove swamp infested with snakes and mosquitoes.

Development was no easy task. The land first had to be cleared. Then new earth was dumped on and around the irregular swampy surface in order to build it up and increase its size. After obtaining permits from state and federal governments, the fill material was dredged from the shallow bottom of Biscayne Bay. This process has added much additional land to the Florida waterfront area.

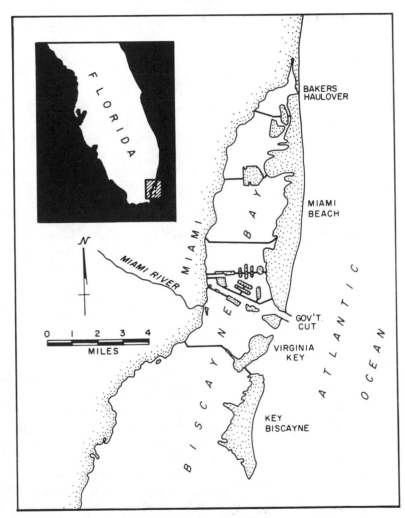

Fig. 24. *Sedimentary barrier islands of Florida's east coast.*

These islands are called the sedimentary barrier islands, first, because of the material of which they are made and, second, because of their location. Sedimentary material is composed of grains of various sizes and types of rock that have been carried from one locality and dropped as sediments in another. The sediments of these islands are made of a mixture of limestone and quartz grains. The elongated north-south position of these islands creates a barrier between the Atlantic and the mainland.

The fact that they contain a good deal of quartz makes them different from the other islands to the south that are made almost exclusively of limestone. Quartz could only have come from areas north of here, where it is part of the bedrock. Longshore currents carried it to the south.

These sediments were dropped in this particular place because of the position and shape of the bedrock that lies beneath them. As shown on the Florida reef tract map (p. 70), the islands can be traced in a line to the south that points directly to Soldier Key and the other Florida Keys. These latter keys are made of an ancient coral reef rock. A low-lying extension of this old reef runs to the north and disappears below sea level in the area opposite Miami. It formed here a low bedrock barrier upon which the sediments were deposited. When the longshore currents carrying the sediments from the north reached the submerged barrier, they lost their velocity and dropped their sediments over and around it. Thus Miami Beach, Virginia Key, and Key Biscayne have a foundation of hard coral reef limestone topped by an accumulation of sedimentary sand deposits.

Virginia Key is unique in that its southeastern tip houses one of the most important clusters of oceanographic institutions in the world. Here are located the Rosenstiel School of Marine and Atmospheric Science of the University of Miami; the Southeast Fishery Center's Miami Laboratory of the National Marine Fisheries Service, part of the National Oceanic and Atmospheric Administration (NOAA); NOAA's Atlantic Oceanographic and Meteorological Laboratories; and the Miami Seaquarium. It would be difficult to find a more strategic location for an oceanographic center. Its mild, semitropical climate makes it possible to carry on marine research throughout the year. Its central location gives easy access to two continents, North America and South America, as well as the wide expanse of the Atlantic Ocean with its many auxiliary bodies of water. Even the Pacific is relatively near, thanks to the Panama Canal.

Key Biscayne is without a doubt one of the most beautiful islands in the southern part of the state. The greater part of this elongated island is government owned and preserved for public use. The northern half, about two miles in length and three-fourths of a mile in width, belongs to Dade County. The southern tip, about one and one-fourth miles in length, has been acquired recently by the state as the site for Cape Florida State Park. In between, there is an attractive residential area of private homes, a limited number of high-rise apartments and hotels, and a well-kept business area. Practically the whole northern and eastern coast is bordered by a sand beach. A strip of tall coconut palms runs for miles down the center of the island and serves as a median for a four-lane highway.

Key Biscayne's Petrified Forest

Key Biscayne's geologic history is very similar to that of the other islands of the chain. It also originated as a sandbar, it was and still is partially covered by mangroves, and it has its share of filled land. Rickenbacker Causeway, which links Key Biscayne to the mainland, crosses over Bear Cut, which separates Key Biscayne from Virginia Key (fig. 25). On the eastern side of the road, at the extreme northern point of Key Biscayne, there is a small parking area that adjoins a swimming beach facing Bear Cut.

This can be the starting point of a fascinating and informative excursion for those who are interested in understanding Key Biscayne's geology. The hike should be made at low tide, and one should wear old shoes, preferably rubber-soled sneakers. Eastward along the sandy, northern shore of the island, about a half mile from the starting point, the sand beach is bordered by a small rock reef (fig. 26), which ex-

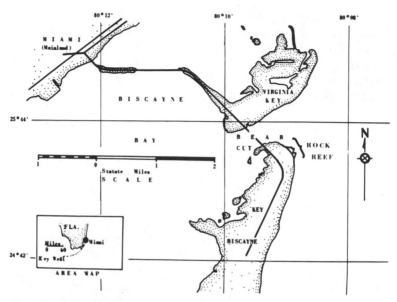

Fig. 25. *Map of Bear Cut area showing Rickenbacker Causeway and rock reef.*

Fig. 26. *View of Key Biscayne rock reef at low tide, looking north.*

tends along the shore for about 425 yards and juts seaward for a maximum distance of 115 yards. It is possible to walk directly on its irregular surface at low tide; it is barely awash at high tide. Figure 27 diagrams the jagged outer edge of the reef.

The surface of the reef contains numerous shallow pits and basins where solution and boring organisms have worn through a crust averaging about three inches in thickness. A vertical view at the edge of the reef reveals an unusual rock formation. Beneath the surface crust, cylindrical rods, averaging about half an inch in diameter, extend downward a few inches to a foot or more (fig. 28). Many are attached at their lower ends to a layer of horizontal rods that can be traced along the exposure for several feet. Other vertical rods extend downward from the horizontal rods to reach a lower layer of the horizontal rods, resulting in an irregular lattice at the outer edge of the reef and a veritable forest of rods inward from the edge.

In places it can be seen that the rods are embedded in a soft, calcareous, and quartzitic sand. Where the sand has been removed by wave erosion, a porous reef rock with large cavities results. The porous rock is limited to the areas that have been affected by wave action; the remainder is a soft sand formation with embedded rods.

This rock reef has been a favorite collecting spot for biologists, botanists, and geologists for many years. Small marine

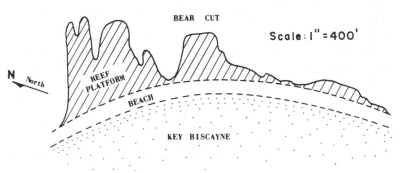

Fig. 27. *Diagrammatic sketch of the mangrove reef, Key Biscayne.*

Fig. 28. *Edge of reef showing lattice structure made by horizontal and vertical rods.*

animals and plants, especially those that inhabit the intertidal zone and cling to the rock surface—barnacles, snails, worms, and red and green algae—are numerous. The origin of the reef has been a subject of much discussion and controversy.

A drop of acid on the surface quickly reveals that the rods are made of limestone—calcium carbonate. Some limestone is made organically, by the action of animals and plants as they precipitate calcium carbonate from the water in which they live. The shells of clams, snails, and the hard parts of corals are examples of this process. Other limestones are made inorganically by the chemical precipitation of calcium carbonate from seawater without the help of living things.

The fact that this limestone is made of rods and that they are arranged in some sort of design and order is a strong indication that it was made organically by animals or plants. One of the most common structures of this shape is made by a certain type of marine worm. These worms live in limestone tubes that they have secreted around themselves. Many worms living close together produce large worm reefs. Examination of the rods of Key Biscayne reef shows that many are hollow and thus they might be the former homes of some tube-building animal. In fact, many of the rods are at present occupied by worms. At first it might appear that the rock reef was formed by tube-secreting worms, but one must question whether the worms may have secreted the rods or merely occupied them after some other organisms had created them.

A factor that makes one skeptical of a worm reef theory is that the arrangement and shape of the rods are different from those of any known worm reef. In most worm reef structures the tubes are very close together and are usually of smaller diameter than those at Key Biscayne.

No lime-secreting animal known produces a structure similar to that of the Biscayne reef. We note, however, that the shape and arrangement of the rods is superficially similar to the roots of a certain type of mangrove. Several species of mangroves are found in the marine swamps of southern Florida. The most common is the red mangrove, *Rhizophora mangle,* which has distinctive "prop roots." The black mangrove, *Avenidia nitida,* is not as common but nevertheless is well represented.

The adult black mangrove has three types of root structures that are important here. At about the level of high tide

Fig. 29. *Living black mangrove showing lateral and vertical roots (pneumatophores).*

or somewhat lower, a ring of lateral roots extends outward horizontally from the trunk, similar to the spokes of a wheel (fig. 29). These may be as much as 15 feet long, tapering slightly toward their tips and branching into two sub-branches at frequent intervals. For most of their length they are embedded under several inches of sand or mud. From these grow vertical roots, called pneumatophores, which extend above the surface of the ground and which probably aid in plant aeration. Extending downward from the lateral roots are other roots known as geotrophic roots, which end in thin branchlets that are used for anchoring and absorbing (fig. 30).

The structural similarities between the roots of a living black mangrove and the shape and arrangement of the rods of the reef are striking. They are identical in size, type, arrangement, and general structure (figs. 31 and 32). It can be seen readily that if many black mangrove trees were growing close

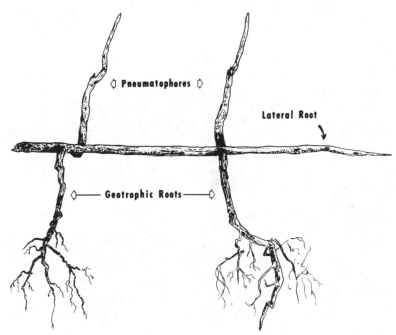

Fig. 30. *Basic root structure of black mangrove.*

Fig. 31. *Black mangrove with roots exposed at low tide.*

Fig. 32. *Edge of reef showing platform and rods after sand has been removed by wave erosion. Note similarities between the roots of the living black mangrove (fig. 31) and the rod structure of the rock reef.*

together, the roots would exhibit the same kind of arrangement as that which is seen at the edge of the rock reef. It would seem plausible that an extensive black mangrove forest once occupied the area. To verify this, the next step is to examine the roots in more detail with the aid of a microscope, comparing a cross section showing the internal structure of one of the roots of a living mangrove with that of a cross section of one of the rods. Figure 33 shows that a living mangrove root is composed of three major parts—an outer layer known as the periderm, an adjoining layer known as the cortex or aerenchyma, and a central section, the vascular area. Each has its own cell structure and purpose. A cross section of a well-preserved fossil rod (fig. 34) is strikingly similar. Although it is made of limestone, it possesses the texture of the woody plant. The hard, outer layer or rim of the rod is equivalent to the periderm of the root. The first inner layer has a remarkable likeness to the aerenchyma of the mangrove root; it has the same general texture and degree of porosity. The second inner layer or central section of the rod is similar to the vascular section of the plant. Like the latter, it is tighter, less porous, and somewhat darker than the aerenchyma. Microscopic examination of thin sections of the internal parts of the rods has failed so far to reveal any cellular structures. Apparently the woody cells of the original plant become obliterated during the process of fossilization. Even so, it seems clear that the rock reef originated as the roots of a forest of mangroves. To determine how long ago the trees were alive the radiocarbon dating method was used. Samples from the calcareous center of one of the rods yielded an age of approximately 1,960 plus or minus 180 years, and from another rod, 1,000 years plus or minus 140. A sample from the sand in which these rods were embedded indicates an age of about 6,000 years. Even though the sand is older by three or more times than the rods, both the sand and the rods are geologically young compared to the 100,000-year-old corals of which the rocks of the Florida Keys are made.

It is difficult to tell how large this mangrove swamp formerly was. At first we could only be certain from the area now exposed by the rock reef. The fact that the sea waves are

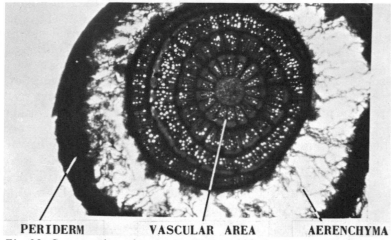

PERIDERM   VASCULAR AREA   AERENCHYMA

Fig. 33. *Cross section of pneumatophore of black mangrove (enlarged).*

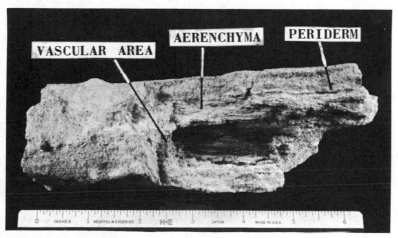

VASCULAR AREA   AERENCHYMA   PERIDERM

Fig. 34. *Fossil lateral root of black mangrove.*

rapidly eroding it, however, makes it clear that at one time it must have extended oceanward considerably beyond its present edge. How far is hard to determine, but a shallow underwater platform seaward of the edge, marking the level of erosion, indicates a distance at least as great as the present maximum width of the rock reef.

The landward edge of the reef is covered by sand, living mangroves, and other plants. Only by a series of borings could one tell how much of the rock reef is buried in this direction. The whole eastern side of Key Biscayne is bordered by one of the best sand beaches in the area, much is now a part of the Crandon Park public bathing beach. The question of whether or not an extension of the ancient mangrove reef lies beneath the sand remained unresolved until a hurricane helped with an answer. In 1965 Hurricane Betsy swept across the island and removed a great deal of the sand. Patches of the old forest were uncovered as far south as the hotel area.

A very rough estimate of the extent of the old mangrove swamp is that it occupied a strip at least 1,000 feet in width in an east-west direction and had a north-south length of about three miles. The western part is now exposed or buried beneath sand while the eastern part is covered by shallow water. So it is evident that the reef rods are remnants of a swamp of black mangroves that at one time extended seaward beyond the present edge of the reef and landward over the northern and eastern part of what is now known as Key Biscayne. The soil in which the roots were embedded was a calcareous-quartzitic sand located between the high and low tide levels. The trees were subsequently destroyed and the woody material of the roots subjected to slow decomposition. At the same time, the woody tissue of the roots was gradually replaced by calcium carbonate. The wood of the roots became petrified, and thus formed what might be called the petrified forest of Key Biscayne.

Petrified wood is commonly found in many places all around the world. Probably the best known area is the petrified forest of Arizona. Although our petrified forest is relatively insignificant in size to this, nevertheless it is a most unusual and perhaps unique example. Mangroves do not adapt easily to fossilization, and this is the only place in the world where such a forest has been reported. Fossil mangroves may very well be located in other places, but so far this is the only area where it is definitely known. For this reason, the little rock reef of Key Biscayne is rather special and should be preserved.

4| Corals and Coral Reefs

The continental shelf is narrowest off the south-eastern part of Florida; however, three of the most interesting and important geologic features of our country occupy this narrow strip. The Florida Keys are located parallel to the edge of the shelf from Soldier Key to Key West (see fig. 39). Seaward of the Keys, between them and the edge of the continental shelf, is a narrow, flattish area where the only living coral reefs of the entire continental United States grow and flourish—the Florida reef tract. Between the Keys and the mainland is a line of shallow bays—from Biscayne Bay through Florida Bay. The geologic history of these three areas is closely entwined, and a knowledge of all three is necessary to understand any one of them. We can refer to them collectively as the coral reef area.

It is fairly well known that the Keys represent a line of ancient and elevated coral reefs, but the only way to arrive at a full understanding of them is to study first the living corals of the reef tract and see what is happening there.

Corals are certainly among the most beautiful living things found in the seas. Although many corals look like plants, they are carnivorous animals that eat other small animals. The term coral is so loosely used and applied to so many animals of such different appearance and structure that it is a difficult group to describe. They all have at least one thing in common, however, they all possess a soft, sacklike or tubular body with one opening or mouth surrounded by numerous tentacles that help them to get their food.

Each body is called a polyp (fig. 35). This places them in a phylum, or major group, known as the Coelenterata, a lowly type of invertebrate slightly higher in the scale of life than the sponges. Well-known examples of coelenterates are jelly-fishes, sea anemones, and Portuguese men-of-war. Some coelenterates remain as single individuals; others form colonies of hundreds or thousands of individuals.

In a general way it can be said that true corals differ from other coelenterates by the fact that they have hard skeletons produced by the soft living tissues of the polyps. Of these there are two main divisions—the semiprecious corals and the stony corals.

Semiprecious Corals

Some corals have been used for many years for jewelry. The most famous is the red coral of the Mediterranean. Its

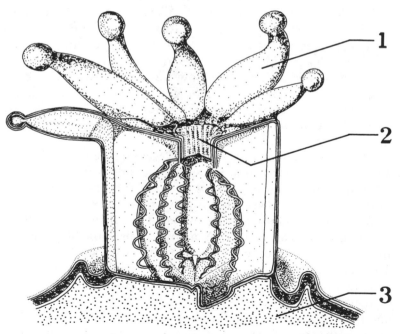

Fig. 35. *Coral polyp. 1: tentacle; 2: mouth; 3: hard skeleton.*

hard skeletal material has a horny texture. When it is brought out of the sea, it is green in color. It dries to shades of deep red to milky pink. It is this coral that is responsible for the color known as coral. Recently red coral has been found off Hawaii, Formosa, and Okinawa in the Pacific. The closely related "black coral," also found in Hawaiian waters, has recently become much sought after for the manufacture of jewelry.

Stony Corals

White or stony coral is by far the most prolific and important of all the corals. It is made of lime carbonate and in this respect is different from the horny material of the semi-precious coral. It is the one that makes our coral reefs—the true reef coral. One of the best known students of coral reefs, C. M. Y. Yonge, after completing a year's study of the Great Barrier Reef of Australia, stated that "corals are the greatest of the world's builders, and the Great Barrier is their most magnificent creation."

Corals displayed for sale in store windows are usually white, but some merchants try to improve on nature and dye them pink, green, or some other color. These corals are merely the skeletons of coral colonies. The soft living parts that once covered the skeletons like thin slimy skins have dried up and been washed away. Each of the small pits or cavities that line the surface of the hard structure was occupied by a living polyp (fig. 36). The tissues of each polyp extracted calcium carbonate from the seawater and precipitated it at its base.

Coral polyps grow in num-

Fig. 36. *Part of a branch of living staghorn coral,* Acropora cervicornis, *showing polyps with tentacles extended.*

ber by budding or division. In this way, hundreds or thousands of polyps create more skeletal material, and the colony slowly grows in size. When something happens to kill the polyps, the colony stops growing. If one lifted a common type of coral out of the water, such as a brain coral (fig. 37), and rubbed his hand over the soft, slimy, living surface, he would remove most of the living organism. That would be the end of the living colony, and only the dead skeleton would remain.

Conditions of Coral Growth

Reef corals are sensitive animals that can live only in warm seawater; a temperature below 60 degrees Fahrenheit is fatal. Also, they are shallow-water animals; you can almost say the shallower, the better. They are seldom found at a depth below 150 feet where sunlight, which is necessary for their growth, is absorbed by the overlying water. The great majority thrive best in the zero to 50 foot zone. While none can live above the level of the sea, at low tide many colonies may

Fig. 37. *Brain coral,* Diploria. *Living tissue covers the surface. The individual polyps are spaced at intervals in the grooves.*

be exposed for several hours and can survive until the next high tide comes in. Corals are sensitive to waters that contain sediment. Their tentacles can ward off only a limited amount of sediment, after which they may smother.

Reef corals support minute single-celled plants within their tissues. These plants provide, through their waste products, some nourishment for the corals, which in turn provide the plants with the carbon dioxide that they must have in order to carry out their photosynthetic activities.

For man, the most important function of reef corals is the production of those vast structures of limestone known as coral reefs. Coral reefs are not by any means made of corals alone. In fact, many other calcareous organisms, both animal and plant, may contribute more to the volume of a reef than do the corals. But corals are the architects and builders of the reef. Living coral polyps establish themselves on a shallow platform and begin to develop as colonies. These grow horizontally and vertically, producing forms characteristic of their species. As they get larger, they grow closer together. The spaces between them become filled in by many kinds of smaller calcareous organisms, both entire skeletons and fragments broken up by wave action, and boring animals and plants. Small pieces of mollusk shells, sea urchins, forams, and calcareous plants are among the most important detrital ingredients of the filler.

The debris settles on the sea bottom and gradually increases in thickness, surrounding and covering the lower parts of the adjacent coral colonies, which continue to grow upward. New colonies may develop between the older ones or on others that have died and therefore have stopped growing. In time, a considerable thickness of coral heads surrounded by finer detrital material will develop. The whole mass may eventually be solidified by chemical action and by the binding action of certain algae into a solid rock mass. The resulting coral reef is a heterogeneous, calcareous structure composed of large and small coral colonies surrounded by a cement of fragmentary particles of many sizes and organic types.

The conditions necessary to coral growth explain the location of living coral reefs today. Geographically they are limited to the tropical and subtropical regions of the earth. Roughly, they occupy the area between 25 degrees North latitude and 25 degrees South latitude, extending around the world. This is the coral reef belt (fig. 38). Since corals can live only in shallow water, they are commonly found around islands and shores of continents. These offer the corals shallow platforms on which to become established. Some, however, are found in areas of the ocean far from land. These are the mysterious reefs called atolls.

Even within the coral reef belt, coral reefs are more prolific in some areas than in others. For example, they are more flourishing along eastern continental shores, which are more favored by warm ocean currents, than western shores. Also, the greatest concentration of reefs is in the Pacific and Indian oceans.

Types of Coral Reefs

For many years coral reefs have been classified under three main types: (1) fringing reefs, (2) barrier reefs, and (3) atolls.

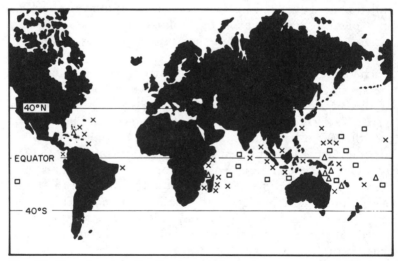

Fig. 38. *Map showing distribution of the major coral reefs throughout the world. X: fringing reefs; △: barrier reefs; □: atolls.*

Fringing reefs develop in the shallow water surrounding an island or mainland. Barrier reefs are separated from the land area by lagoons, areas of relatively protected water. Atolls, found in the open ocean at some distance from a land area, are composed of a roughly circular reef that surrounds a central lagoon. Natural passageways that cut through these reefs in places provide connections between the open ocean and lagoons.

In addition to these main types, at least two other terms are used to describe reefs in more detail. In places where there is a broad reef, such as our Florida reef tract, it has been convenient to speak of the seaward edge as the "outer reef." Behind the outer reef, on the reef platform, small reefs frequently rise to sea level. These are the so-called patch reefs, which grow in the back reef zone, back of the outer reef.

5| *The Florida Reef Tract*

The Florida reef tract is an arc-shaped shallow water limestone shelf that is bounded on the seaward side by the deeper waters of the Straits of Florida and shoreward by the Florida Keys (fig. 39). It has a length of about 150 miles and an average width of four miles. The surface is irregular and varies in depth from zero to 50 feet. A glance at the hydrographic chart (fig. 40) shows a pattern of low ridges or banks alternating with valleys or channels that run parallel to the Keys and the outer edge of the shelf. The shallowest parts

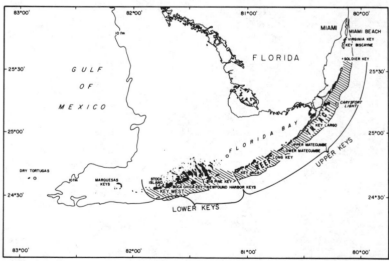

Fig. 39. *The Florida reef tract.*

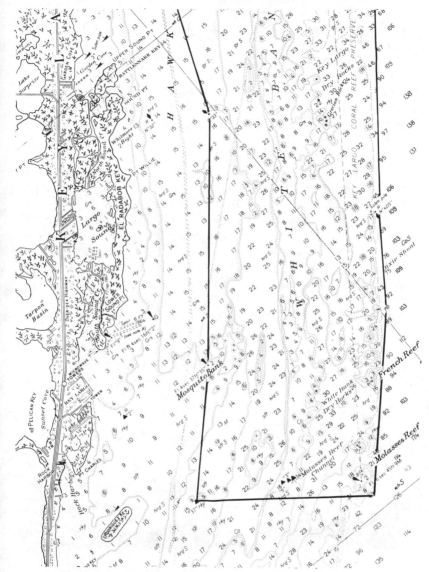

Fig. 40. *Section of a hydrographic chart of the reef tract illustrating channels, banks, patch reefs, and various outer reefs such as French Reef. Hawk Channel, White Bank, and the patch reef at Mosquito Bank are good examples. The area of John Pennekamp Coral Reef State Park is shown by darker lines.*

hug the Keys and have a width to seaward of a mile or more. In general, the deeper valleys occupy the central area and alternate with parallel ridges. The outer edge of the platform is very irregular due to the intermittent presence of coral reefs.

The dominant pattern is made more complex and irregular by the presence of small reef patches that dot the surface in many places. These have a tendency to arrange themselves in lines parallel to the long direction of the tract. Many are located on the banks, but others are scattered irregularly over the platform (fig. 41). The reef tract is no place to run a boat without a good chart, and even with one, a careful lookout must be kept for suddenly appearing coral heads.

Two main physiographic forms rise from the bottom of the reef tract or platform: ridges or banks, and coral reefs. The former are mainly composed of loose particles of limestone—sedimentary detritus of many kinds of invertebrate marine animals and plants. These have been carried by numerous currents over the tract and deposited as the elongated banks previously described. A good example is White Bank,

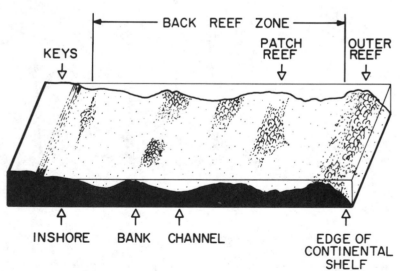

Fig. 41. *Diagrammatic east-west cross section of the reef tract.*

which parallels Key Largo for a distance of 30 miles. Hawk Channel is an excellent example of one of the channels.

Coral reefs produce the hard, resistant forms scattered in many places over the tract. They commonly rise to sea level or slightly below and are divided into two types, depending upon their geographic position: (1) outer reefs, which form at the seaward edge of the platform; (2) patch reefs, which are found behind the outer reefs in the back reef zone. The patch reefs are clusters of relatively small, irregularly-shaped coral masses that rise like knolls from the surface of the platform. In places where a few are close together they may occupy an area of 5,000 square feet or more.

Good examples of outer reefs are at Carysfort Lighthouse and Key Largo Dry Rocks. Mosquito Bank, within the boundary of the John Pennekamp Coral Reef State Park, is a good example of a patch reef in the back reef zone.

There is no doubt that in Florida the best developed coral reefs are on the northern section of the tract. Those farther south are not as flourishing. Evidently the environment in the south is not as favorable for the growth of corals. A main reason for this difference is that in the northern section Key Largo acts as a solid, more or less continuous barrier between the reef tract and the Bay environment landward of the Keys. It thus prevents the sediments of the Bay area from encroaching on the coral reef territory. In contrast, farther south the many channels that connect Florida Bay to the reef tract permit a relatively free flow between the two subenvironments. Alligator Reef, which lies opposite the wide channel that separates Upper Matecumbe from Lower Matecumbe, is an example of what the presence of sand and silt can do to prevent flourishing coral growth.

Hurricanes can also do great damage to coral reefs. French Reef, which undoubtedly was at one time a flourishing coral community, has been changed by the destructive force of hurricanes into a barren high point on the outer reef.

Corals of the Florida Reef Tract

Corals are the most attractive and appealing features of the Florida reef tract. Coral colonies vary greatly in size and shape. Some are branching, with thin, delicate branches or with thick, massive armlike structures. Others are rounded, with irregular or subspherical shapes. Some are small, a few inches across, while others stand eight feet high and have a diameter of 10 or more feet.

The surfaces of certain colonies are pitted by small, circular depressions (fig. 42). Each depression is divided into smaller pie-shaped parts by vertical partitions or septa that join in the center and give it a star-shaped appearance. Such corals have commonly been named "star corals" (fig. 42). Each depression is occupied by a separate polyp during the life of the colony. Other colonies possess meandering grooves

Fig. 42. *Star coral,* Montastrea cavernosa *(approximately x 10). A polyp occupies each depression.*

over their surfaces. The living polyps are spaced at fairly regular intervals in these grooves. Such corals are known as brain corals (fig. 37). Both types are common in the reef tract.

It has been estimated that about 50 species of corals are now living on the Florida reef tract. These are the same species as those commonly found in the entire Bahamian and West Indian coral regions. Three main environments affect coral growth on the Florida reef tract: (1) the outer reef at the seaward edge of the platform, (2) the patch reef in the back reef zone, and (3) the inshore shoals. Many species of corals are able to live in any of these three locations; others flourish in one or more of the environments but are scarcely able to eke out an existence in others. Some few species can live only in one of the environments. Attempts to transplant them to other locations have failed.

Corals of the Outer Reef

The corals of the outer reef live in the most active environment. The full force of the waves from the open ocean, carrying potential food, strikes the outer edge of the platform.

Fig. 43. *Large branching fronds of the moosehorn coral,* Acropora palmata, *surrounded by a thicket of* Acropora cervicornis *at upper portion of outer reef about 5 miles southwest of Carysfort Light, Florida reef tract.*

The greatest variety of corals can be found living here. A few, however, dominate and provide the main structure for reef building. Two of the most beautiful species on the tract are the moosehorn coral, *Acropora palmata,* and the staghorn coral, *Acropora cervicornis.* In spite of the fact that they possess long branches, and therefore are susceptible to wave damage, they grow in such large numbers that they produce veritable coral thickets (fig. 43). As the branches grow closer to each other, they are cemented together by encrusting calcareous algae to form a coalescing structure. Other branches are broken up and produce fragments of debris of all sizes. These accumulate between the coral colonies and are also cemented together to form a massive wave-resistant limestone reef. In addition to the species mentioned above, the star coral *Montastrea annularis* (fig. 44), and several species of brain corals, *Diploria,* are common. Good places to observe these corals are at Carysfort Lighthouse and at Key Largo Dry Rocks.

Fig. 44. Montastrea annularis, *star coral, with multilobular heads. Large branches of* Acropora palmata *in foreground.*

Corals of the Back Reef

In the back reef quieter water prevails because of the protection afforded by the outer reef. The back reef is also subjected to more silt and sand constantly on the move by the active currents of the tract. An examination of a typical patch reef in the back reef area shows a decided difference as compared to those of the outer reef (fig. 45). This does not mean the species are different. In fact, most of the species are

Fig. 45. *Example of corals found on patch reefs of the back reef area. Patch reef community at Mosquito Bank consisting of* Montastrea annularis *(M)*, Porites asteroides *(P)*, Halimedia opuntia *(H), and several acylonarians (A) such as sea whips and sea fans* (Gorgonia flabellum).

common to both types of reef. There is a great difference, however, in the number of colonies of the main species.

While in the outer reef the moosehorn coral and staghorn coral are the dominant species and the chief architects in the formation of the reef, in the patch reefs their place is taken by the star coral *Montastrea annularis.* Large colonies up to 10 feet in diameter are more abundant here than any other species. Most of these are subhemispherical in shape and others are lobate. Staghorn coral is present, but it is not nearly as conspicuous. Moosehorn coral is conspicuous by its absence.

Other prominent corals in the patch reefs are *Porites astreoides,* which possesses very small depressions for its polyps, and several species of the brain coral. Mosquito Bank is an excellent place to examine the corals of the patch reefs.

Corals of the Inshore Shoals

Areas of the reef tract that hug the Keys, and where the water is 10 feet or less in depth, are occupied by a very distinctive assemblage of coral species. On shoals in this environment four species are common. The star corals *Solenastrea hyades* and *Siderastrea radians* occupy the rocky bottoms of the shoals, while the rose coral *Manicina areolata* and the small finger coral *Porites divaricata* inhabit the sandy bottom.

They are all characterized by their small size. What they lack in this respect is made up by their numbers. There are large areas where colonies of the finger coral *Porites divaricata* are so numerous and so closely spaced that it would be difficult to step between them (fig. 46). One of the best places to observe the shoal corals is Rodriguez Bank, which is located in the shoal area of Key Largo about five miles north of its southern end. No large colonies similar to those that occupy the patch or outer reefs can be found in this inshore zone.

The distribution of coral species thus is not indiscriminate, but is controlled by environment, with each of the three major environmental zones possessing characteristics that are favorable to some species and unfavorable to others.

Fig. 46. *An example of a species found on inshore shoals,* Porites divari-cata, *finger coral.*

Growth Rate of Corals of the Reef Tract

How fast corals grow depends on their species and the environment in which they live. My research included a study of the growth rates of some of the common species of corals on the reef tract. After they were carefully measured, weighed, and photographed, small living coral colonies were cemented to concrete tiles. They were then attached to the sea floor by means of an iron stake that penetrated a hole in the center of each tile (fig. 47). The tiles were removed from the sea bottom and the corals remeasured at six-month intervals to see what changes had taken place.

This study lasted for a period of three years. It was determined that the average upward growth of specimens of the star coral *Montastrea annularis* (fig. 48) amounted to 10.7

Fig. 47. *Growth rate experiment. The tiles with small coral colonies cemented to them were stationed on the sea bottom at a depth of about 10 to 15 feet. The bottom here is strewn with branches of the staghorn coral,* Acropora cervicornis. *The planting was made next to a cable to make it easier to locate on the sea floor.*

millimeters. At this rate an upward growth of one foot would take about 28.5 years. Some other well-known corals of the reef tract grow at a much faster rate (fig. 49). For example, E. A. Shinn (1966) determined that the common staghorn coral, *Acropora cervicornis,* has an annual growth rate of 100 millimeters or about four inches a year. Thus it would take only three years for a colony to grow a foot higher, considerably faster than the more solid *M. annularis.* Shinn discovered that most corals do not grow at an even rate. Following periods of relative quiescence, their growth rate spurts. The reason for this is not definitely known, but it may be due to seasonal variations of water temperature.

Besides corals, many other important lime-producing animals and plants contribute to the growth of the Florida reef tract, but in this discussion it has been my intent to focus on only the most essential ingredients in the geological story of this region.

The only way to understand and really appreciate Florida's coral reef area is to see it. Those who live near or who have the opportunity to visit the region should drive to the Keys

Fig. 48. *Star coral,* Montastrea annularis, *is brought up from the sea floor for remeasurement.*

Fig. 49. *Staghorn coral,* Acropora cervicornis, *is raised from the sea floor for remeasurement. The scale in the background is marked in inches.*

and take a ride in a glass bottom boat at the John Pennekamp Coral Reef State Park. This unique underseas park of 75,000 acres is the first underwater park in the continental United States. It was established to preserve the corals and associated animals and plants of the only living reef formation along the coast of the continental United States. Its land base, located off U.S. 1 on Key Largo, is easy to reach.

6| The Florida Keys

The story of the Florida reef tract can now be used in understanding the next and very important segment of this history—the making of the Florida Keys. This fascinating area is different from any other part of our country, and the only place where one can drive over stepping-stone islands made of old coral reefs for a distance of over 100 miles into the open ocean and see some of the most colorful seascapes of the world.

The Florida Keys are a narrow chain of small islands extending from Soldier Key on the north to Key West on the south and west, a total distance of about 150 miles (fig. 51). Together they describe a graceful arc with the convex side bounded by the Atlantic Ocean and the concave side by Florida Bay, with its adjoining bodies of inland water, and the Gulf of Mexico.

The greater part of the Keys is low-lying, projecting only two to four feet above high tide. The highest point is on Windley Key, which reaches an elevation of 18 feet. Other high islands are Lignum Vitae, Plantation Key, Key Largo, and Big Pine Key. Mangroves line the edges of many of the Keys. In fact, it is estimated that mangrove swamps occupy at least one half of the total area of the Keys, thus giving the impression that the land is much larger in area than it really is.

Sand beaches are remarkably few and are of limited extent. Those that occur are found chiefly on the Atlantic side, notably on parts of Key Largo, Plantation Key, the Mate-

Fig. 50. *Aerial view of the Florida Keys from Upper Matecumbe Key looking northeast, with the Atlantic Ocean on the right and Florida Bay on the left.*

cumbes, Long Key, and Key Vaca. Intertidal flats border both sides of the islands. These are shallow water areas, barely covered at low tide, which gradually slope into the deeper water of the surrounding platform. One of the best examples is found on the ocean side of Spanish Harbor Key, located just northeast of Big Pine Key.

The surfaces of most of the Keys are flat and of about the same elevation throughout their extent. Others, particularly the northern Keys, are a few feet higher along their central elongated axis and slope gradually to the ocean and bay sides. Although bedrock extends only a few feet above the high tide level, there are numerous outcrops, the best of which have been created artificially. Small quarries and borrow pits, dug during the construction of roads, and boat slips expose the bedrock in many places both inland and along the shores. Offshore dredging operations that involve making new land and converting mangrove areas into habitable sites are numerous. The rocks thus exposed contributed a great deal to the knowledge of the terrain and reveal much that was not available to earlier investigators.

The Keys can be divided naturally into two groups—the upper keys and the lower keys. The upper keys extend from Soldier Key to Big Pine Key, a distance of 110 miles; the lower keys extend from Big Pine Key to Key West, a distance of 40 miles. There is a great deal of difference between the two groups, but many people take the trip to Key West without realizing this. One of the major differences can be recognized easily with the aid of a map (fig. 51). The upper keys are oriented along the line of the arc in a southwesterly direction. The lower keys are each oriented almost at right angles to the arc and are separated from each other by narrow channels that run in a northwest-southeast direction.

The Rocks of the Keys

Like the southern part of the mainland, all the rocks of the Keys are made of limestone. It has long been known that two main types of limestone are present here. One of these is an

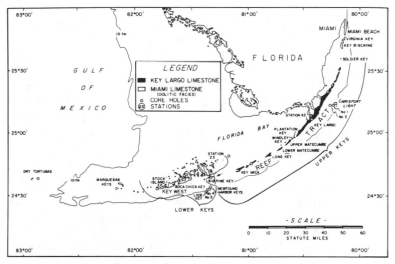

Fig. 51. *Main features and geology of the Florida Keys.*

ancient coral reef rock known as the Key Largo limestone; the other is the oolitic facies of the Miami limestone. Figure 51 shows the location of each of these. It can be seen that the upper keys are made of Key Largo limestone and the lower keys of Miami limestone. This is one of the reasons why there is such a difference between the two groups of keys—they are made of different types of rock and so were created under different conditions.

The Upper Keys

If one wishes to see real coral rock, the place to visit is the upper keys. There are numerous places where excellent exposures display the fossil coral reef and spectacular views of the main corals that produced it. The Key Largo limestone, which has been called the backbone of the Keys, extends on the surface from Soldier Key in the north through the westernmost of the small Newfound Harbor Keys (south of Big Pine Key), where it terminates at Newfound Harbor Channel, a distance of 110 miles. The Key Largo limestone extends over a much greater territory than its surface outcrops indicate. In the north, borings have revealed its presence under Miami Beach. There is little doubt that it continues south beneath Virginia Key and Key Biscayne in a direct line to Soldier Key. Although no surface exposures of the Key Largo limestone are visible west of Big Pine Key, dredging projects and a core boring at Key West indicate that it underlies all of the lower keys, proving that in this area the coral reef limestone is covered by the oolitic Miami limestone.

An extension of the arc of the Keys in a westerly direction from Key West leads to the Marquesas Keys, Rebecca Shoal, and the Dry Tortugas. A core recently placed at Dry Tortugas encountered the Key Largo limestone at a depth of 30 feet below sea level, thus the formation is known to exist at a distance of 70 miles beyond Key West. Here, then, is a coral reef that, at the time it was living, had a total length of at least 220 miles. This is a large reef, not as large as the Great Barrier Reef of Australia, but still a very sizable one.

One next wonders whether this reef is a thin layer of coral reef limestone deposited over some other kind of rock or if it is of considerable thickness. In other words, did it exist for a relatively brief time or did it last, geologically, for a very long time? Again, the only way to answer this question is to drill and examine the cores brought to the surface. A core taken near the northern tip of Key Largo showed the Key Largo limestone to be 145 feet in thickness. Ten miles farther south on the same key it thins to 75 feet. At Grassy Key its thickness increases to 170 feet. At Big Pine the core drill went through 20 feet of oolite followed by 170 feet of Key Largo limestone without reaching its base. A well drilled at Key West penetrates 27 feet of oolite, followed by 170 feet of the Key Largo limestone, without reaching its base. It appears that the formation is of variable thickness, that it rests upon an undulating, irregular surface, and that it must have been in existence many hundreds of years in order to have built a thickness of this magnitude.

There are many places where the Key Largo limestone can be closely examined in the upper keys. Probably the best exposures occur in the quarry on Windley Key (fig. 52). Here

Fig. 52. *Exposure of Key Largo limestone in quarry wall at Windley Key showing irregular distribution of coral colonies.*

it is possible to walk along quarry sides, which are as high as 10 feet, for a distance of about 500 feet. At first glance the general impression will be that of a hodgepodge of corals of all sizes, kinds, and shapes, scattered without plan and separated from each other by hardened detrital material of sandgrain size and larger. This is typical of coral reefs. They are not layered or bedded deposits, such as ordinary sedimentary rocks. They are a mixed-up, disorderly formation. This is, however, just what one would expect to find in an ancient coral reef after seeing the living corals and deposits of the living coral reef tract.

A closer examination shows that some of the coral colonies are much larger than the others and more numerous. Many of these stand in an upright position just as they stood when they were alive. Some are on their sides as though they had toppled over. The larger corals formed the framework of the reef. Existing close together, they created a bafflelike structure that trapped smaller particles so that together a solid mass was created. What is seen here is exactly what is happening today on the living reef tract and is also comparable to what is happening in mangrove swamps that trap sediments washed in by wave action.

A real coral reef, such as this one, is a deposit that has been formed in place or *in situ*. The large coral heads, many of which are standing in an upright position, could not possibly have been grown at some other locality and transported to this one.

Corals of the Key Largo Limestone

The corals and their arrangement found in the quarry are typical of those found throughout the long extent of the Key Largo limestone area. The corals that form the framework of the Key Largo reef—the real architects of the structure—are the star coral *Montastrea annularis* (hemispherical and branching) and several species of brain coral, *Diploria* and *Porites astreoides*. Smaller corals such as *Agaricia agaricites* and the finger coral *Porites porites* fill in the interstices along with fragments of all types. In a few places in the Key Largo limestone, small thickets of *Acropora cervicornis* are conspicuous. *Monastrea annularis,* however, is without question the dominant coral of the formation.

Most of the corals that are common on the living coral reef tract can also be found in the ancient Key Largo reef. There is a greater similarity, however, between the Key Largo corals and those of the patch reefs than with those of the outer reef. One very significant difference between those of the Key Largo and those of the outer reef is that the very important moosehorn coral, *Acropora palmata,* has never been found in the Key Largo formation of the Keys.

In addition to the quarry, there are several other excellent places to view the Key Largo formation. One of the most striking is on Key Largo at Station 62 shown on figure 51. Here the Cross Key Canal has been cut through the key from Florida Bay to the Atlantic Ocean. This cut, with a maximum exposure of about 10 feet in height and a length of 675 yards, reveals a magnificent view of the upper part of the old coral reef. This is an easily accessible place that every visitor to the Keys should see. The best view can be obtained by walking east toward the Atlantic along the northern side of the cut.

One of the striking features of this exposure is the presence in the cliff face of the opposite side of several large colonies of the star coral, *Montastrea annularis,* in growth position. These possess the multilobate or branching form. The machine that cut the canal bisected the colonies and exposed their internal makeup. Many of the individual columnar branches were sliced through the middle. One of the best colonies is shown in figure 53. The rod measures six feet in length. The maximum width of the colony is about eight feet, and the maximum height is seven feet. The branches or lobes average about four inches in diameter and are separated from each other by about three-quarters of an inch. It can be easily seen that colonies of this type are excellent baffles for trapping sediment carried by currents over the platforms on which they grow.

Most persons who see a coral like this are interested in knowing how long it took for the colony to grow to this size. The length of time necessary to produce such a colony can

Fig. 53. *Section of Key Largo limestone at Cross Key Canal on Key Largo showing large colony of* Montastrea annularis. *The rod markings are at intervals of one foot.*

best be determined by measuring the longest branch that shows continuous growth. It has already been mentioned that branches of *Montastrea annularis* grow at the rate of about one foot in 28.5 years. Assuming that the upward growth rate of the fossil coral was similar to the growth rate of the living *M. annularis,* it is estimated that it took about 230 years for this colony to grow to its maximum height of seven feet. This coral lived approximately 100,000 years ago.

The age of an individual colony is one thing, but it is also interesting to try to estimate how long it took to build an equal thickness (seven feet) of the reef. Certainly it must have taken somewhat longer for the accumulation of the surrounding detritus. A rough estimate of the total time taken for the accumulation of seven feet of the Key Largo limestone at this place is between 250 and 500 years.

Other good places on Key Largo to see what the Key

Largo limestone looks like are at a development known as Marion Park, located a few miles south of the Cross Key Canal on the ocean side, and at Seven Acres, still farther south, on the Bay side of U.S. 1. In both places boat slips cut into the rock offer excellent views.

The Lower Keys

Driving south across the bridge from Spanish Harbor Key to Big Pine Key, one arrives in a new and very different geologic region. Instead of the Key Largo coral reef limestone, one now drives over oolite rock of the Miami limestone. Also, instead of following the long axis of each key, you ride at right angles to the axis and therefore over the narrowest parts. The numerous bridges between islands cross over narrow channels that run in a northwest-southeast direction.

Big Pine is the island where the Key Largo and the Miami limestone meet. The best place to see the meeting point between the two formations is the point called Station 23 in figure 51. Here the oolite overlaps the old coral reef at a low angle in a southern direction. The oolite covers all of the lower keys and is thinnest over their southern borders, increasing in thickness to the north. It reaches its greatest known thickness of about 35 feet on the northern part of Stock Island, which adjoins Key West.

The rocks of the lower keys undoubtedly began as an underwater east-west mound of unstable oolite lying north of, and parallel to, the Key Largo coral reef. As it grew higher by the addition of more ooids, the mound gradually became broader to the south as well as to the north and eventually covered the corals. The shape and size of the original mound can be determined easily by drawing on a map a pencil line around the area of the oolite islands. The result is a sketch of an east-west oolite deposit about 40 miles in length and 10 to 15 miles wide. If we could have flown over it in that early day, it would have looked very much like the present unstable oolite ridge of the Great Bahama Bank (fig. 19). The only difference is that the two mounds are oriented in different directions.

Observing the oolitic mound of the Bahama Bank will help explain the orientation of the individual keys and the channels that lie between them. Tidal currents cut channels in the unstable oolite at right angles to the long direction of the mound. When the sea level lowered during the glacial times, the exposed oolitic material hardened, as in the case of the Atlantic coastal ridge. The subsequent rise in sea level enabled the ocean waves to attack the mound. These waves concentrated on the old tidal channels originally formed while the mound was composed of loose ooids and eventually formed the narrow channels that today separate the lower keys from each other. This process continues at the present time, and gives evidence that the present shape and orientation of the lower keys had their start in the underwater topography created by tidal currents on an unstable oolitic mound.

The Origin of the Upper Keys

Although we have seen that the lower keys were formed of oolitic Miami limestone, the origin of the upper keys—the real coral reef keys—is quite different. They are made of Key Largo limestone, which also extends beneath the Miami limestone of the lower keys. It is easy enough to say that these keys were made of ancient coral in shallow water and that later the sea level lowered leaving the reef high and dry. This is, however, only part of the story.

We know now that our living coral reef tract or platform has at least two types of reefs on it: an outer reef near the edge of the continental shelf and patch reefs in the back reef area behind the outer reef. The question is to which of these two types did the Florida Keys belong? At the time they were living, were they an outer reef or were they patch reefs? If they were outer reefs they must have been exposed directly to the waves of the open ocean, and there could have been no shallow platform seaward of them. If they were a line of patch reefs they must have been established in the back reef area of an old, higher platform that could have extended seaward as far as the present reef tract. In that case, the platform presumably would have been edged on the seaward side by an outer reef. What clues can we find to arrive at a solution to the problem?

Certainly the most important clues of origin should come from the composition of the old reef itself. If it was an outer reef, it should reveal the characteristics found on the present-day outer reef. If it was formed in what was a back reef environment, it should possess features similar to those found in the numerous patch reefs of the present back reef region.

The coral species of the Key Largo limestone of the Keys are commonly found in the living patch reefs. All flourish best in a somewhat restricted environment with limited wave energy. It has been shown that the most common and the most important of the main frame-building corals of both the Keys and the patch reefs is the star coral, *Montastrea annularis*. The main frame-building corals of the outer reef are the moosehorn coral, *Acropora palmata*, and the staghorn coral, *A. cervicornis*.

There are certain species that are found in typical outer reefs in the Florida and West Indian regions that so far have never been found in the Key Largo limestone of the Keys. One of these is *Acropora palmata*. This is among the best-known and most widely dispersed species of corals living in the Atlantic today. It thrives where the surf is strong. It is never found on the patch reefs in the back reef area where wave action is restricted. It seems obvious, therefore, that the Keys were originally formed as patch reefs in a back environment and were never associated with the strong wave action common to outer reefs.

If they were patch reefs, they must have grown in the back reef area of a platform that extended some distance seaward, possibly as far as the present platform. This old platform must also have been made of old Key Largo limestone and must have been edged by an outer reef that had corals characteristic of this zone such as *Acropora palmata*.

Now we wonder what has happened to the reefs of this old platform. We at least can see some of them—the present Keys. But how about the patch reefs and the outer reef that stood seaward of the Keys? It seems that the Keys are all that remain. What happened to the others?

The sequence of events that have led to the present condition is shown by means of block diagrams. The condition across the coral reef tract in Key Largo time is shown in figure 54. The platform itself is made of Key Largo limestone. The corals that now compose the Florida Keys are shown as living patch reefs in the back reef area. In front of them are two other lines of patch reefs and, finally, an outer reef at the seaward edge. At that time the sea level was about 25 feet higher than at present. This took place about 100,000 years ago, during the Sangamon interglacial period. Then came a time when the climate changed and it became colder due to the accumulation of great thicknesses of snow and ice in the polar regions (the Wisconsin glacial period). The water that produced this snow and ice came from the evaporation of sea water, and as a result, the level of the sea lowered slowly but considerably. Areas that had been covered by water now became dry land. The water over the old Key Largo platform was drained off, and the reefs were exposed. Wave action at the edge of the platform attacked the outer reefs and the platform behind them until much was eroded away. Erosion continued as the sea level gradually lowered until the upper 50 feet or so of the seaward edge was removed. Less erosion occurred farther inland as wave action

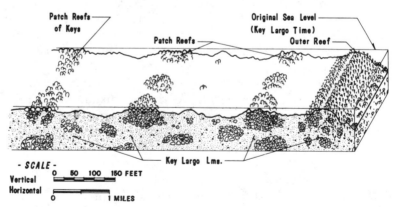

Fig. 54. *Generalized block diagram showing conditions across the Florida coral reef tract in Key Largo time.*

was not as violent here. However, the patch reefs of the central part of the platform were cut away. Only the lower part of the patch reefs, which are now the Keys, remained.

The surface of the platform at the time of maximum erosion is shown by the irregular line in figure 55. Years later the climate again began to change in the direction of greater warmth. As the ice in the northern regions melted, the sea level slowly rose. When the water reached the level of the old eroded platform, new coral reefs were developed, and new sediments were deposited on the platform. An outer reef grew at the seaward edge, and patch reefs grew in the back reef area to the landward. Finally, the sea reached its present level, and the conditions of the present reef tract prevailed. The patch reefs of the Florida Keys still remain above the waters of the sea—all that are left.

In order to substantiate this theory of the back reef origin of the Keys, one other piece of evidence was needed—evidence to prove that there was an outer reef in existence at the time the corals of the Keys were living. This was obtained by coring through the present outer reef down to the eroded surface and into the old Key Largo limestone. This was done at Looe Key, opposite Big Pine Key. Here, at a depth of

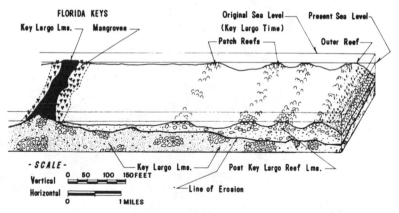

Fig. 55. *Generalized block diagram showing conditions across the Florida coral reef tract in the present time.*

about 60 feet, specimens of moosehorn coral, *Acropora palmata,* were found. This was the first time that this species had ever been reported in the Key Largo limestone. It demonstrates the existence of an outer reef at the time the keys were being made as patch reefs in the back reef area.

7| Bays of the East Coast

The southern tip of the Florida mainland is bathed on the east and south by several shallow bays, the northernmost of which is Biscayne Bay. South of this are the smaller bays of Card Sound and Barnes Sound. Two narrow, irregular sandbanks encrusted by mangroves separate them from the much larger Florida Bay, which extends around the southern tip of Florida and merges into the Gulf of Mexico. Biscayne Bay is about 30 miles in length with a maximum width of eight miles. The combined length of Card Sound and Barnes Sound is about 12 miles; their average width is three and one-half miles. They are bordered on their eastern side by the sedimentary barrier islands, a shallow tidal bar known as the Safety Valve, and the northernmost Florida Keys. Taken collectively, the average maximum depth of the three bays is about 10 feet.

Most people who cruise on any body of water are interested in knowing what the surface of the bottom is like—whether it is flat or rough with high and low spots. In most cases their interest is confined to knowing where the high places are—where the water is shallow—as the safety of the boat depends on it. The existence of buoys and other markers makes this rather easy. Hydrographic charts recording numerous soundings help a great deal, but even these give us only a sketchy idea of what the bottom really looks like.

The best information concerning the topographic features of an underwater floor can be gotten from a bathymetric map showing the surface by contour lines, which indicate depth beneath sea level. No accurate bathymetric map of the three bays under discussion has been made, but figure 56 shows the approximate location of the 6, 9, and 12 foot contours. From this it can be seen that the northern part of Biscayne Bay near Miami Beach is very shallow, averaging less than six feet, and that it becomes somewhat deeper to the

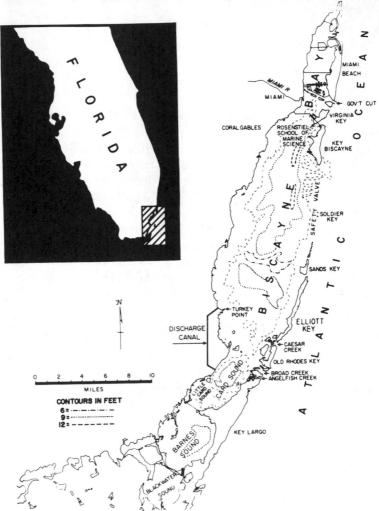

Fig. 56. *Bathymetric map of the Biscayne Bay area.*

south. As the Bay broadens to the south, the map shows that a broad, shallow shelf less than six feet deep occupies the western side and gradually increases in depth eastward to about 15 feet just west of Key Biscayne. This is near the middle of the relatively deep north-south axis of the Bay, which extends from Virginia Key to Elliott Key. South of this, Featherbed Bank, an east-west trending shoal that is nearly awash at low tide, divides the Bay into a northern and a southern section. The deeper axis of the southern section lies slightly east of the central part of the Bay and continues to Card Sound, where the depth is less than six feet.

Both Card Sound and Barnes Sound are shallow, basin-shaped depressions with maximum depths of slightly more than nine feet in the center of each. It can be seen that the deepest part of Biscayne Bay is the elongated north-south depression that hugs closely to the islands that rim the eastern side and hold the water within the basin.

The several gaps between the islands permit an exchange of water with the Atlantic. One of these is between Virginia Key and Key Biscayne. This narrow gap, Bear Cut, is 12 to 15 feet deep. A much broader one, the so-called Safety Valve, lies between the southern tip of Key Biscayne and Soldier Key. This is a mud bar cut by a number of tide channels 10 to 20 feet in depth. Two other deep channels are found farther south, one at the north end (Caesar's Creek) and the other at the south end (Angelfish Creek) of Old Rhodes Key. These are about 25 feet deep.

The floors of these bays have a very thin cover of unconsolidated sedimentary material over hard bedrock. Over most of the area the thickness of the sediment is less than six inches. Thicker deposits are located in the northernmost part of Biscayne Bay between Miami Beach and the mainland. South of this, deposits reaching as much as 40 inches in thickness are found along the eastern side of the Bay, in general following the area of its deep axis.

By means of the information obtained from probing and coring, it is possible to determine the thickness of the sediment and also to acquire a good deal of information concerning the shape and topography of the underlying bedrock surface. If we could entirely remove all the unconsolidated material, the hard bedrock surface would look much like the surface shown on figure 57. This figure shows an elongated trough, the northern end shaped like the prow of a canoe. The eastern side of the trough is made of the coral reef Key Largo limestone. The latter is relatively low in the north in

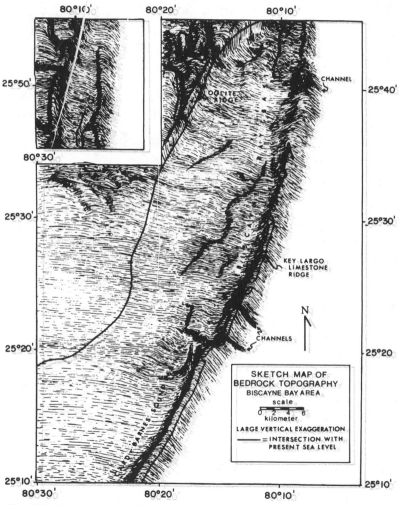

Fig. 57. *Sketch map of the bedrock topography of the Biscayne Bay area.*

the vicinity of Miami Beach, becoming higher toward the south until it is seen as Soldier Key, Elliott Key, and other emerged keys today.

The northwestern side of the trough is made of the oolitic facies of the Miami limestone, part of the Atlantic coastal ridge. Farther south, since the ridge trends to the west, the bedrock shoreline loses its clifflike character and becomes a nearly flat surface sloping very gently toward the deeper part of the trough.

The bottom area of the trough, between the two sides, is characterized by two basins, the Biscayne Bay basin and the Card-Barnes Sounds basin. They are separated from each other by a shallow platform east of Old Rhodes Key except for a deep axis that runs the entire length of both basins. This axis adjoins the western slope of the coral reef Key Largo ridge. The depths of the basins are from nine to slightly more than 18 feet in the deep axis.

There is much evidence that at one time this old bedrock surface was dry land exposed to the atmosphere. Valleys cut by streams can be seen along the eastern side of the trough, and in at least three places streams from the trough have cut channels (such as Caesar's Creek Channel) through the Key Largo ridge. In addition, the bottom is pitted with solution holes caused by fresh water dissolving the limestone. This happens today on the land wherever there is limestone.

There is no doubt that the entire floor of the Bay is made of hard limestone, on the east side, the Key Largo limestone and on the west, the Miami limestone. As these come together beneath the Bay, it is believed that the Miami limestone extends eastward over somewhat more of the Bay bottom than the Key Largo limestone from the east. The contact line between the two formations cannot be accurately drawn.

The thin cover of unconsolidated sediment that overlies the bedrock formation of the bays is composed of several very different kinds of sediment distributed rather irregularly over the bottom:

(1) Lime mud and sands are a common part of the sediments, especially in the central body of the bays. They have been derived from the limestone rocks surrounding them.

(2) Quartz sands of two types have been found in many

sediment accumulations in the Bay. Since the local rocks contain only a minute amount of quartz, they must have come from some other source. In one type the quartz grains are angular in shape and clear in appearance (fig. 58). In the other type they are more rounded and clouded (fig. 59). The first type occurs as accumulations of pure quartz, while the other is a mixture of quartz and carbonate grains. The two types are concentrated in different sections of the Bay. The pure quartz deposits are found in the northern end, especially between the mainland and Miami Beach. The angular condition of the grains indicates that they have been transported only a relatively short distance from their source. The quartz-carbonate type is located along the eastern side of the Bay on both sides of the barrier islands. The worn and cloudy appearance of the grains shows that they have been transported a considerable distance and are the longshore deposits that have been the chief contributors to the barrier islands.

(3) Freshwater calcitic muds lie directly on the bedrock in the shallow water along the western shore of the bays.

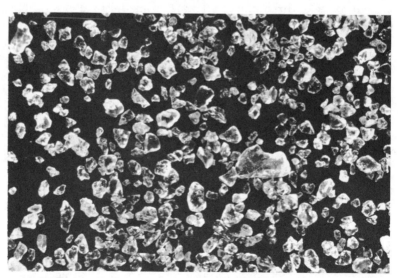

Fig. 58. *Quartz grains, angular and clear in appearance, from bottom sediments in northern Biscayne Bay.*

Fig. 59. *Quartz grains, rounded and clouded, mixed with calcareous grains, carried and deposited here by longshore currents from the north.*

They rise to sea level from the maximum depth of about six feet.

(4) Fibrous peat deposits, which may be five feet or so in thickness, are found on top and merging with the calcitic muds.

So far, examination reveals the following information concerning the bays: they are basin-shaped, with the long axis running in a northeast-southwest direction; the bottom and sides of each basin are made of hard limestone—the eastern side, the Key Largo coral reef rock—the western, the oolitic Miami limestone; and overlying the hard bedrock of the bottom are various thin deposits of soft sediments.

These facts lead to an understanding of the geologic story of the bays. It has been previously shown that the Keys and the Miami limestone were both formed when the sea level was about 25 feet higher than it is today. Picture therefore what this area was like at that time, about 100,000 years ago. On the east side the corals of Elliott Key, Key Largo, et cetera, were living, and the reefs reached to nearly sea level.

On the western side the unconsolidated mound of oolite that was to become our Atlantic coastal ridge also was built up to near sea level. Between them was the depression or lagoon that was to become our bays.

The fact that corals were living certainly means that the climate was warm, very much like that of today. This was the time of the warm Sangamon interglacial period. Then a change in climate occurred and a new glacial period, the Wisconsin, was ushered in. The accumulation of huge amounts of snow and ice in the polar regions resulted in a slow lowering of the sea level. Gradually the Key Largo reef was entirely exposed as was the oolite mound. With exposure to the atmosphere, the limestone deposits of both became hardened into rock.

The lowering of the sea level was not continuous. There was at least one interval of standstill at what is today about eight feet above sea level. At this level the sea waves pounding against the Miami limestone cut the notch that can still be seen at Silver Bluff in Miami. During the lowering of the sea level, pure, angular, quartz sands were carried from the emerging land to the north and deposited in the northern part of the lagoon west of Miami Beach. This was accompanied by an increase in erosion. During this time the seaward edge of the coral reef tract was being attacked by wave action and its surface lowered. The water in the lagoon was drained off, and eventually the bottom of the lagoon was entirely exposed. Then streams forming on the surface created valleys, some of which have been preserved. Sinkholes and other forms of limestone solution also developed. It is believed that the sea level eventually may have reached a maximum low level of about 300 feet beneath the present surface.

About 20,000 years ago, milder climatic conditions heralded the end of the Wisconsin glacial period, and the sea level began its slow rise. When it approached the level of the reef platform, new coral reefs were developed on the eroded surface of the pre-Wisconsin platform. This development continued to the present, and then resulted in the formation of the reef tract as it is today. Farther to the west, at the same time, the rising sea water reached the lagoon behind the Key Largo reef, and Biscayne Bay and the other bays were born. Some of the deposits now found in the bays reveal some of the conditions here at that time. The presence of some fresh-water calcitic muds lying directly on the bedrock six feet and less below present sea level indicates the existence of a fresh-water body in the depression before the saltwater encroach-ment. Immediately above the calcitic muds and in many other places over the bedrock surface, thick mangrove peat is found today. Since mangroves can grow in salty waters, this shows the gradual encroachment of marine water as the sea level rose. If one could have had a view of the bays at that time, he would probably have seen them mainly as mangrove swamps.

As the sea level continued to rise and the bays deepened, other unconsolidated sediments, including lime muds, were deposited. At this time the sedimentary barrier islands also started to form as the longshore sediment, consisting of a mixture of carbonate and quartz sands, began dropping on the eroded and submerged Key Largo ridge. This sediment, as mentioned earlier, made of worn shell fragments and rounded, cloudy quartz grains, indicating considerable abra-sive action, is similar to the sands along beaches farther north. It is thus believed that the sand that formed the bar-rier islands was brought by longshore currents from some distance to the north.

Finally, the postglacial deposits were subjected to consid-erable current action within the bays, distributing them irreg-ularly over the bottom. This accounts for the thin coating of sands and weeds over large parts of these bays, especially near the center, and the much greater thickness along their sides.

8| Florida Bay

Florida Bay is roughly triangular in shape, with the sharp apex at the point where the mainland and the Keys nearly meet. It is bordered on the north by the Florida mainland, on the south by the Florida Keys, and opens on the west to the Gulf of Mexico. It is approximately 850 square miles in area.

Florida Bay is very shallow, with a few points reaching a maximum depth of eight or nine feet. Over the central part, its average depth is between four and five feet. The most striking feature of the topography at its bottom is the presence of a winding, or anastomosing, arrangement of elongated mudbanks that join at irregular intervals to create a lacelike pattern (fig. 64).

The lower areas surrounded by the banks are known as lakes; they average about four to six feet in depth. The banks rise to near sea level and at times are exposed during low tide. Mangroves and other brackish water plants have attached themselves to the banks, helping to hold them in place. In some areas it is evident that storm action has built the banks a little higher than the level of the water, creating small islands or keys.

Foundation of Florida Bay

Florida Bay is underlain, apparently over most of its area, by the hard Miami limestone (oolitic facies). On top of the nearly flat surface of the Miami limestone, there lies a coating

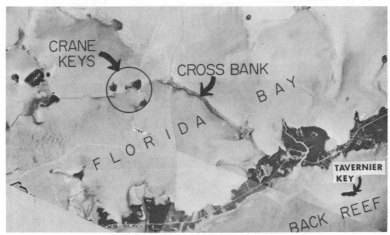

Fig. 60. *Aerial photograph of a section of Florida Bay north of Tavernier Key showing mud lakes, banks, bars, and mounds.*

of unconsolidated, fine-grained, calcareous mud. This varies in thickness from a few feet near the mainland to as much as 10 or 12 feet near Cape Sable. The average thickness of the mud over the banks is three feet and over the lakes about six inches. The sediment coating the Miami limestone is very young and was deposited in postglacial times. Radiocarbon dates indicate that deposition began about 4,000 years ago. Since the Miami limestone is about 100,000 years old, the difference in age between it and the overlying calcareous mud is considerable (fig. 60).

About 90 percent of the unconsolidated sediments in the Bay are made of calcium carbonate. When one considers the large area of the Bay, the fact that it is covered by from six to 10 feet of lime sediment, and the fact that this took place within a period of 4,000 years, it is evident that Florida Bay manufactures a tremendous amount of limestone.

Limestone is one of the main rocks of the earth's crust, not only because there is so much of it (about 20 percent by volume of all the sedimentary rocks), but also because of its intrinsic value. For example, a great deal of oil is found in limestone formations all over the world. Consequently, in the search for oil it is very important for geologists to understand as much as possible about the different kinds of limestone and their origins. If they can see the chemical, biological, and environmental conditions under which various kinds of limestone are now being made, they can apply this knowledge to an understanding of those ancient limestones which are potential reservoirs of oil.

The waters around southern Florida produce more limestone than any other part of our country. Actually it is a region of several limestone factories, with each one producing a different kind of limestone under different conditions. Coral limestones are made on the reef tract, oolitic limestones on the nearby Bahama Banks, still another type in Florida Bay.

Geologists from all over the world come to this great natural limestone manufacturing center to see and study limestones in the making. Since limestones are sedimentary rocks, the best way to understand their origin is to carefully examine the particles of which they are composed. If, for example, the particles are made of fragments of corals and associated organisms, it is easy to determine that the rock was made under tropical, saltwater conditions at shallow depths.

Some limestones have particles that are large enough to be easily identified by the naked eye. Other limestones may be finer grained and the help of a microscope is needed for identification. Such particles may be the size of sand grains or larger. Still others are so fine that even a high-powered microscope is of little help; these are the most difficult rocks to interpret. The shells of lime-secreting animals that possibly produced them may have been smashed by waves and wind action over such a long period of time that they have been ground into a silt or powder so fine that all traces of the original parent material have been eliminated. Or they may have been formed by inorganic means—by chemical precipitation without the aid of any type of marine life. Since rocks of this fine-grained character are very widespread, their origin has been the subject of much study.

Origin of the Florida Bay Sediments

Florida Bay is one of the main laboratories for this study. The reason for this is the presence of the fine-grained sediments that have accumulated and are still accumulating on the floor of the Bay. The size of the particles is the same as that of the fine-grained rocks that have been so baffling to interpret. If it could be revealed how these fine-grained muds are now being produced, it would help greatly to solve the mystery of the origin of the widespread fine-grained limestone rocks produced in the past.

This may seem to be a rather simple thing to do, but actually it has been a rather complicated problem. Some researchers have thought they were made inorganically; others suggested that bacterial action was involved; and still others thought that the disintegration of shelled organisms, either animal or plant, was the key. The shell-secreting animals and plants now living in the Bay include mollusks—pelecypods (bivalves), gastropods (snails), and small one-celled animals known as foraminifera. At least 100 genera and 140 species of mollusks have been reported. In a few places near the passages that connect the Bay to the reef tract, some corals typical of shoal environments are found. That the shells of these lime-producing animals have been the source of the lime sediments of the Bay is possible; there is no doubt that they have contributed a certain amount, as their fragments have been recognized in particles of sand and even in silt. This does not account, however, for the very fine mud particles that make up the major part of the sediments. It is hard to believe that the shells were also responsible for these. Florida Bay is a lagoonlike body with relatively quiet water. There are no breakers or strong wave actions similar to those found on the outer part of the reef tract. The Bay is protected by the Florida Keys, and it is difficult to conceive the existence of a grinding and smashing action that would be necessary to reduce shells to a fine-grained mud condition. If the larger shells are not the main source of the sediments, then we must ask if there are any lime-producing organisms that secrete skeletons made of very fine particles.

The floor of Florida Bay is covered with seaweeds. These are mainly lowly plants known as algae. Figure 61 shows a specimen of *Penicillus,* one of the green algae commonly found here. Each plant is composed of a rootlike structure known as a holdfast, a stem or stalk, and a head or crown

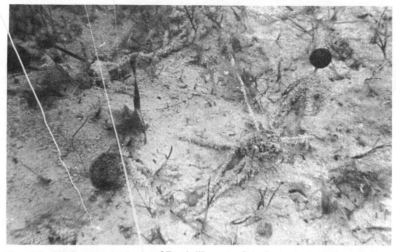

Fig. 61. *Underwater view* of Penicillus *sp. showing vegetative reproduction. The rhizomes that connect these plants were exposed by waving the sediment away.*

made of filaments (fig. 62). *Penicillus* has the ability to secrete very fragile skeletons composed of tiny crystals of calcium carbonate known as aragonite. The stem and the filaments of the head are enclosed in a delicate sheath of aragonite crystals. When the plant dies, the woody part decomposes, the crystalline sheath disintegrates, and all that is left are the tiny clay- or mud-sized crystals (fig. 63). Here, then, is a possible source of lime. The only question is whether the algae could produce a sufficient quantity of these microscopic crystals to account for the extensive deposits of lime mud on the floor of the Bay during the estimated required time of 4,000 years.

A group of Miami geologists attempted to solve the problem directly by trying to determine the amount of minute calcareous crystals that were formed by the algae in one year (Stockman, Ginsberg, and Shinn, 1967). They established stations in a variety of localities. At each station, stakes were driven in the Bay bottom so that one or more square plots, each measuring 0.5 meters on a side, were outlined. The number of *Penicillus* plants in each plot was recorded. About

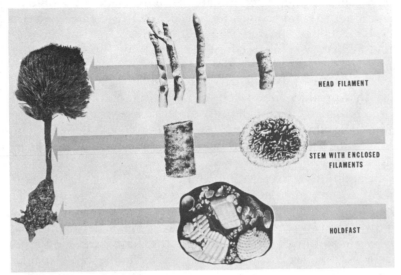

Fig. 62. *Morphology of* Penicillus *sp.*

HEAD FILAMENT

STEM WITH ENCLOSED FILAMENTS

HOLDFAST

Fig. 63. *Microscopic crystals of aragonite from* Penicillus *(greatly enlarged).*

every 20 days for a year, each of the stations was visited and the changes noted. Some plants had died and new plants had taken their places. It was determined that the life span of a plant was about 40 days— thus there would be about nine generations per year. At the end of a year's observation, the number of plants that grew, died, and contributed their skeletons was recorded for each of the plots.

The geologists needed next to determine the weight of calcareous crystals contributed by an average plant. It was estimated that each square meter over the area was occupied by two plants. The estimate of the life span, the average abundance of plants, and the average weight of calcium carbonate per plant made it possible to determine the annual production of mud by *Penicillus*. The results obtained showed that at the present rate of production, *Penicillus* could account for about one-third of the fine calcareous mud in Florida Bay.

Penicillus is only one of several similar algae living in the Bay that produce mud-sized calcareous grains. The scientists agreed that if the skeletons of these were added to the amount produced by *Penicillus*, the total would be sufficient to account for the total production of limestone sediments in Florida Bay.

In addition to the algal lime mud that has originated exclusively within the Bay itself, there is considerable evidence that some comes from outside sources. For example, it is known that currents from the Gulf of Mexico, which enter the wide southwestern opening of the Bay, carry fine mud particles that are deposited to help form the broad banks south of Cape Sable.

Bathymetry and Geologic Development

The unusual honeycomb pattern of the bathymetry of Florida Bay has attracted considerable interest, and its origin has been a subject of much discussion among geologists. Here is a shallow basin with a hard bedrock bottom made of Miami limestone, which slopes very slightly from the mainland to the Keys. Upon the rock foundation is an accumulation, several feet thick, of a fine lime mud. Under ordinary conditions of sedimentation, the fine-grained mud would have been deposited layer upon layer in an orderly fashion, and the result would have been a Florida Bay with a flattish bottom. Instead of that, something has happened to cause the mud and other sediment to form a honeycomb of ridges surrounding shallow lakes.

The fact that there are two different rock deposits, each of very different composition and age, indicates that there are two chapters in the history of the Bay. The older or first chapter deals with the conditions that existed about 100,000 years ago. At this time the Florida Keys were being formed and the coral reef was alive and flourishing. The sea at that time stood about 25 feet higher than at present, and the region of what is now Florida Bay was a lagoon between the living coral reef and the mainland to the north. In this lagoon the deposits of unconsolidated oolite limestone that eventually covered the entire floor were being made.

Then the climate became colder and the sea level lowered as ice accumulated in large masses in the Arctic and Antarctic regions. The whole area, the mainland, Florida Bay, the Keys, and the reef tract became dry land as the sea level dropped well over 100 feet. During this time the newly exposed land was subjected to considerable erosion, and the unconsolidated oolite became hardened to form the Miami limestone.

Then the sea level began its slow upward rise as the climate grew warmer and more water returned to the oceans. If, a short time previous to 4,000 years ago, we could have looked at what is today the bottom of Florida Bay, we would have seen an area very similar to our Everglades. In other words, the Everglades at that time extended over the mainland and as far south as the Keys. It was a low-lying region, sloping gently southward, covered with Everglades vegetation, dotted with freshwater lakes and streaked with rills and sloughs. It was a freshwater environment with no marine life, plant or animal, present.

Finally, about 4,000 years ago the sea level reached the lowest parts of the land, and marine waters began to spread over the low-lying area of what is today Florida Bay. Much of the water first reached the exposed Bay bottom through the channels that separate the Keys. The sedimentary cover and the topography of the Bay as we find them today are the result of the slow encroachment of marine water over a gently sloping freshwater surface that was very much like what we find now in the Everglades. The environmental change from fresh to marine water brought marine animals and plants into the region. Prominent among these were mangroves, calcareous algae, and marine mollusks.

Mangroves played an important part in the early invasion of marine life. The seawater reached the low-lying rill valleys first and gradually extended northward along their channels. The latter became the home for the earliest mangrove devel-

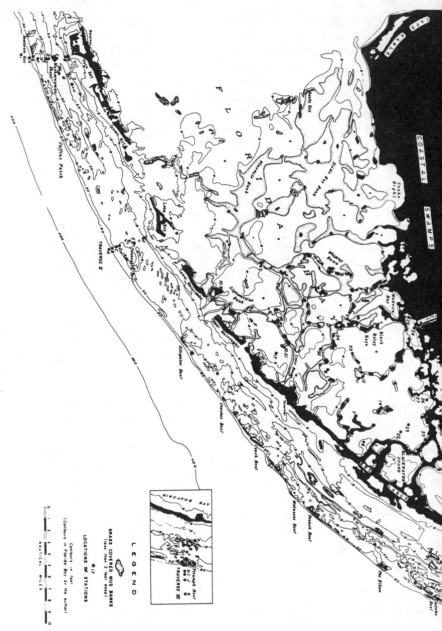

Fig. 64. *Florida Bay and reef tract showing honeycomb pattern.*

opment. Accompanying them were the associated marine grasses. Together they formed the baffles that captured any sediment that came their way. This included the fine-grained lime muds from *Penicillus* and other green algae, which were accumulating in large amounts as the sea level rose, as well as deposits of peat formed from the disintegration of mangroves. The areas of the north-south oriented rills were turned into banks as they became somewhat higher than the surrounding territory.

As the mangrove growths in the rills extended northward into the freshwater zone of the mainland, the southern parts were gradually surrounded by saline waters. As the water deepened and the Bay extended over the land, other banks were formed along its northern shoreline perpendicular to the rill banks. The crossing of these two sets of mounds created an irregular lattice pattern—the spaces becoming the lakes.

Tidal currents and storm waves created larger lakes by the destruction of some of the separating walls. The early formation of the banks and lakes can best be seen along the northern shore of Florida Bay (fig. 64). Here the lakes are smaller and more numerous. The structure farther south, where the lakes are fewer and larger, is the result of marine and tidal current action destroying some of the separating banks.

Tidal currents have helped to control the overaccumulation of lime muds on the lake bottoms and also to increase the height of the interlake walls. In many places storm waves have heaped up enough sediment to create small mangrove islets.

The geologic story of Florida Bay has now been briefly summarized. The two most distinctive features are, of course, the origin of the fine-grained calcareous muds and the unusual topographic features of the Bay floor.

9| *Florida's West Coast*

The western coast of the southern tip of Florida has not received as much attention from geologists as has the more accessible eastern coast. In addition to the more difficult terrain, this is an area where exposures of bedrock are very scarce. It is mainly a place of unconsolidated sediments thickly coated with mangroves and associated plant growth. It is a region where salt water extends for several miles inland before freshwater areas are encountered. Although this region is lacking in exposures of bedrock, it nevertheless has a most interesting story to tell.

The distance from the southern tip northward to Cape Romano is about 50 miles (fig. 65). Extending several miles inland from the irregular and island-studded shoreline lies one of the largest coastal swamps known. Its total area is estimated to be about 200 square miles. When one makes a tour of the region, he can imagine himself in the midst of a prehistoric swamp similar to that which produced our coal deposits. The trees are different, but the general conditions are probably much the same. It is also the area of the Ten Thousand Islands (fig. 66).

The bedrock underlying the southern section is the Miami limestone (bryozoan facies). North of this it is the Tamiami formation of Miocene age. In both areas the bedrock slopes very gently from the mainland into the Gulf of Mexico. An examination shows that lying on the submerged bedrock are unconsolidated sediments of varying thickness, averaging about 12 to 15 feet. The component particles of the sediments throughout the swamp are the same—calcium carbon-

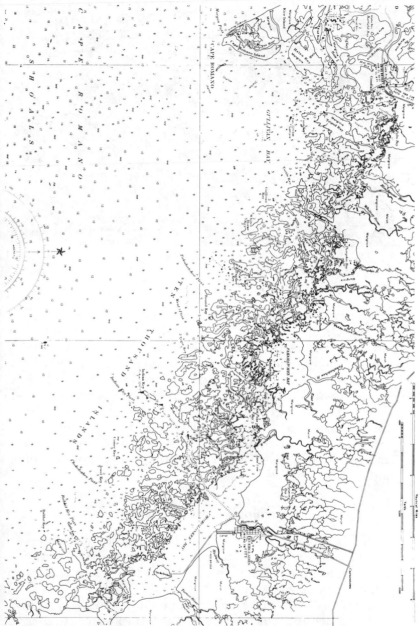

Fig. 65. *Hydrographic chart of the Ten Thousand Islands area.*

Fig. 66. *Aerial photograph of the Ten Thousand Islands showing how the direction of development of oyster bars controls the elongated and sinuous direction of island growth.*

ate, quartz, and organic material. These are distributed in different proportions in different localities. The southern part contains much calcium carbonate and a small amount of quartz. In a northern direction, calcium carbonate particles gradually decrease, and quartz becomes dominant. This is because Cape Romano at the northern end is made almost entirely of pure quartz sand. From this source longshore currents have carried quartz grains southward and deposited them offshore in amounts decreasing with the distance. The southern end is limited to the production of lime materials.

The Geologic Work of Mangroves

This is a region where the geologic role of mangroves is superbly illustrated. These plants perform two functions: they increase the amount of land, and they record and interpret changes in sea level.

Longshore currents coming from the north carry quartz sand and deposit it farther south in deeper water parallel to the shore of the mainland. These offshore deposits are built up by the accretion of more sand and the acquisition of an extensive growth of oysters (*Crassostrea virginica*). The latter cover the sand deposits in favorable places in large numbers, and their limestone shells add to the upward growth of the sediments. These eventually reach the intertidal zone.

Oyster bars, 20 feet or so in width, are exposed at low tide (fig. 67). They can be seen today to grow horizontally in sinuous patterns with branches extending at intervals from the main line of growth. The direction of growth is determined by the location of actively flowing water, for which the oysters have a preference. They grow best at right angles to the tidal currents in order to obtain the maximum amount of nutriment available. In places, the branches merge and form small lakes and bays that become traps for mud and sand.

When the intertidal zone is reached, mangroves become established on and around the oyster beds, eventually replacing them and extending beyond them (fig. 68). In this way the islands of the Ten Thousand Islands archipelago were created. Eventually the tidal currents become so restricted due to the accumulation of more and more islands that the oyster growth declines. With the extensive growth of man-

Fig. 67. *Oyster beds (foreground) at low tide, Ten Thousand Islands. Mangroves in background are encroaching over oyster beds.*

Fig. 68. *Encroachment of mangroves over oyster beds.*

groves, especially the red mangrove (*Rhizophora mangle*), peat is formed from the accumulation of fibrous organic debris of generations of trees.

The Ten Thousand Islands thus began from offshore mounds of quartz sand deposited by longshore currents from the north. Upon these mounds, extensive oyster beds were formed in favorable places. When the mounds were built up to the intertidal zone, mangroves established themselves and islands were formed. Thick deposits of peat have been created by the disintegration of generations of mangroves.

Mangroves as Indicators of Sea Level Change

The surface of the sea is constantly rising and falling. There are times when these changes are relatively rapid and other times when they are so slow they seem to be nonexistent. Sea-level changes can be detected in a number of ways. If, for instance, we can see a place where there is a notch cut into a rock cliff several feet above present sea level, it very likely means that it was cut by wave action at a time when the sea stood at the level of the cut. In other words, the sea level has dropped.

Former higher stands of the sea are generally easier to recognize than ancient lower levels, since the sea, as it rises, covers up the evidence. There are several ways of establishing the sea rise, however, and one of the most interesting is by understanding the pattern of mangrove growth. As we look at the mangrove-lined shore of southwestern Florida, we can tell whether it was produced during a rising sea level or one that has fallen by the amount of peat that has been produced from the disintegration of previous stands of mangroves. In places peat has been found to a depth of 12 to 15 feet.

Mangroves are tidal zone plants. They are born, grow, and die within the limits of low and high tide. Therefore the thickest possible accumulation of peat under standstill conditions of the sea could not be much more than about three feet, which is the tidal range in this area. If then when 12 feet of peat are found beneath a stand of living mangroves, it is clear that this stand could not have been formed while the sea was at its present level. Therefore the sea level must have risen during the time the peat was accumulating. As the sea level rose, generation after generation of mangrove trees deposited their woody substance to make the fibrous peat formations now present. John H. Davis, Jr. (1940) estimated a rise of sea level of about seven feet based on a 10-foot thickness of peat. More recently, D. W. Sholl and M. Stuiver (1967, 1969) have suggested the rise of sea level at about 12 feet. It is believed that it has taken from 4,000 to 5,000 years for this change to occur.

10| *Water Resources of South Florida*

Southeastern Florida, which includes Dade, Broward, and Palm Beach counties, is not only a well-populated region, but it is one of the fastest growing sections of the country. At the present rate of growth it has been estimated that by the year 2000 the population will be doubled. A population of this magnitude requires a great deal of freshwater. Where does it come from? Southern Florida is almost surrounded by plenty of salt water, but as yet methods have not been devised to desalt seawater economically in the magnitude that would be required for all needs.

The rocks in some regions of the United States are so solid and tight that rainwater cannot seep into the ground, and therefore it runs off at the surface and much is wasted. Here southeastern Florida is unique and also fortunate. In Dade, Broward, and Palm Beach counties the bedrock is very porous, and rainwater disappears rapidly beneath the surface. Here if one waters his lawn, the water does not stand on the surface, but sinks from sight quickly no matter how much water is used. Because of this, the chief source of potable water here is rainfall, which collects in the rocks beneath the surface.

The bedrock of southeastern Florida is not only porous with much open space, but it is also permeable, that is, it allows liquids to move easily through it from one place to another. When water first disappears beneath the surface, it moves vertically downward through the rock. When it reaches the level below which all the pore spaces are filled with water, the rock is said to be saturated. This level is called the water table; the water below it, called groundwater, moves chiefly laterally instead of vertically. The water table rises and falls depending on the local rainfall. During a dry period the water table declines. When a dry period ends and rainfall increases, the water table rises.

Early settlers in South Florida soon found that there was a great deal of water beneath the surface. It seemed that wherever wells were dug or drilled, an abundance of freshwater was reached. Even when the wells were pumped for long periods, water continued to be available. Of course, in order to obtain a perennial supply of water, the bottom of a well must be located below the water table (fig. 69). During times of drought, the water table may become so low that the well becomes dry. Also, the elevation of the water table varies from place to place. Usually it is higher beneath hills than beneath valleys; in other words, the water table conforms in a subdued way to the shape of the land.

The Biscayne Aquifer

Over the years geologists and engineers have carried out detailed studies of the water supply so that today a good deal

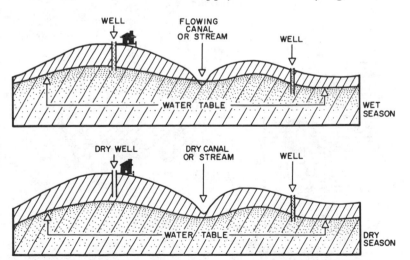

Fig. 69. *Cross section showing the effect of wet and dry seasons on the level of the water table.*

is known about the groundwater resources of South Florida. Chiefly by means of extensive drilling, it was found that a very large area covering all of Dade County, much of Broward and Monroe counties, and the southern part of Palm Beach County is immediately underlain by rocks that contain tremendous amounts of fresh groundwater. The total rock mass that contains the water is known as an aquifer. The shallow water-bearing rocks of southeastern Florida are collectively called the Biscayne aquifer (fig. 70).

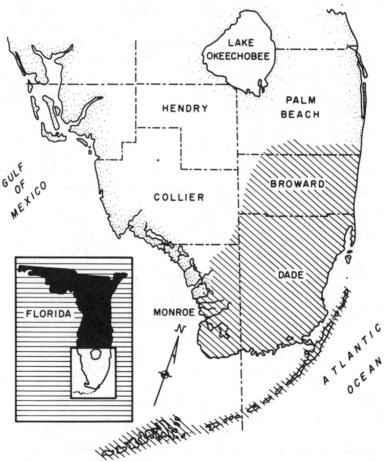

Fig. 70. *Parallel lines show the geographic extent of the Biscayne aquifer.*

Geologists have determined that the Biscayne aquifer is composed of several formations, each of which possesses the properties of an aquifer. Beginning with the topmost or youngest formation, the Biscayne aquifer is comprised of the following formations:

(1) Miami limestone:

Good aquifer.

Contains much pore space and large solution holes produced by dissolving of limestone by underground water.

Chiefly composed of oolitic and bryozoan limestone.

(2) Anastasia formation:

Fair to good aquifer.

Chiefly composed of coquina.

(3) Key Largo limestone:

Excellent aquifer.

Represented on the mainland only as thin layers extending from the Florida Keys.

Chiefly composed of coralline limestone.

(4) Fort Thompson formation:

Undoubtedly the best aquifer in the group.

Yields great quantities of groundwater.

Chiefly composed of solution-riddled marine and freshwater limestones.

(5) Tamiami formation:

Only the upper part is permeable.

The lower part of the formation is relatively impermeable and therefore forms the base of the aquifer (called the aquiclude).

Chiefly composed of shelly limestone in the upper part and sandy or clayey sand in the lower part.

The Biscayne aquifer varies considerably in thickness. In western Broward and Dade counties it varies in thickness from a featheredge to 10 feet and thickens toward the east coast, where it reaches a maximum thickness of about 200 feet in the Fort Lauderdale area. It is 150 feet thick at Miami. Southward from Miami the aquifer gradually thins. At Homestead the aquifer is only 60 feet thick.

The two best water-bearing formations in the aquifer are the Miami limestone and the underlying Fort Thompson formation. Of these the Fort Thompson is by far the more important because it is the thickest and most permeable formation (fig. 71). It is composed of interbedded marine and freshwater limestones. It is very fossiliferous and in places has a "shell-hash" or coquinalike structure. It contains numerous cavities with interconnecting passages. Because of this the Biscayne aquifer is generally regarded as one of the most permeable in the world.

The main source of water in the Biscayne aquifer is local rainfall. The average amount of rainfall per year over the entire area of the aquifer has been estimated to be about 60 inches. The supply in the aquifer is constantly being diminished through the discharge of groundwater into the Atlantic Ocean and Biscayne Bay by the numerous canals that run through the area. In addition, considerable loss occurs by evapotranspiration and pumping.

In 1951 geologists estimated that of the 60 inches of annual rainfall in the coastal ridge area of Dade County, 22 inches were lost by evapotranspiration and surface runoff without reaching the water table. Of the remaining 38 inches that reached the water table, 20 were discharged to the sea as groundwater outflow and 18 inches were lost by pumping from wells and by evapotranspiration from the water table. Since 1950 approximately four inches yearly have been discharged by pumping from wells.

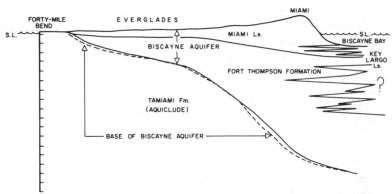

Fig. 71. *Cross section of the Biscayne aquifer showing position and thickness of the main water-bearing formations. Each scale mark indicates 10 feet.*

Artesian Aquifers

Aquifers that are confined above and below by an impervious layer or aquiclude are called artesian aquifers. If rain falls on high land such as that illustrated at the left in figure 72, it would enter the porous aquifer in the area where it has been exposed by erosion. It then percolates downslope by gravity toward discharge areas at the lower levels. Groundwater within the aquifer is contained by the impervious layers above and below. When a well is drilled through the upper impervious layer, the water, now under artesian pressure, rises toward the level of intake. In some wells the pressure is strong enough to cause the water to rise above the surface and gush out with considerable force.

It is evident that the conditions outlined above are very different from those described for the Biscayne aquifer. The latter is a nonartesian aquifer. It is open to the atmosphere and has no confining or impervious layer above it. It does, however, rest on an impervious layer, the Tamiami limestone, which prevents downward seepage. Wells need only to be drilled below the water table in the Biscayne aquifer in order to reach a potable supply of freshwater.

Some artesian water is available in southeastern Florida; the only difficulty is that it is not good, potable water. It comes from aquifers that are geologically older and therefore underlie those of the Biscayne aquifer. Rainwater enters the artesian aquifer in the higher land to the north in the vicinity of Ocala.

One artesian well in Dade County is at Grossman Hammock State Park. Here the water issues out with force, but the well contains a good deal of sulphur and is not potable.

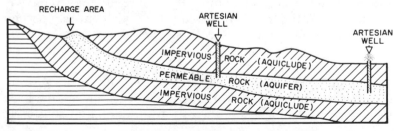

Fig. 72. *Cross section showing structure of artesian wells.*

Some water from artesian sources is being used in small de-salting plants, and there is a possibility that this may provide another source of water supply in the future. There is also a possibility that some surplus water from the Biscayne aquifer might be pumped into one of the deeper aquifers where it could be stored and later pumped to the surface in times of need.

Surface Water Supplies

The water needs of a growing population, an extensive agricultural program, and important natural environments such as the Everglades National Park create intricate problems for those who are charged with the responsibility of maintaining an adequate and equitable water supply. Demands for irrigation and potable water are taken care of by groundwater from the aquifer, but this is not sufficient to supply all needs. The natural environment depends on surface water.

Frequently there is either too much water or too little. During times of abundant rainfall, there is danger of flood damage, especially to the agricultural and urban areas. If preventive action is taken by improved drainage to the ocean, great care must be taken that enough water is saved to supply demands for the subsequent dry period. In order to do this the Central and Southern Flood Control District, working with the U.S. Corps of Engineers and the U.S. Geological Survey, developed a system of storage areas and controlled canals for the region. The center and heart of the system is Lake Okeechobee.

Lake Okeechobee, situated in the northern part of South Florida, has an area of about 700 square miles. After Lake Michigan it is the largest freshwater lake lying entirely within the confines of the United States. The lake is shallow, with an average depth of about 12 feet and a maximum depth of 15 feet.

The chief source of Lake Okeechobee water is the Kissimmee River, which enters from the north. In addition, the lake receives considerable water directly from rainfall over its very large surface. It was previously thought that a great deal of Lake Okeechobee's water came from groundwater seepage from the rock formations surrounding it, particularly from the north. It has since been shown that only a negligible amount reaches the lake from this source.

The natural outflow of the lake in the past was through the wide Everglades to the south. During a dry period a few years prior to 1920, extensive areas of the Everglades near the southern shore of the lake were drained for agricultural purposes. During periods of normal or greater rainfall, considerable damage occurred to the farms when water overflowed from Lake Okeechobee. In order to protect the lands, levees were constructed in the early 1920s along the southern and eastern sides of the lake. This was the first step in utilizing the lake as a tremendous reservoir. Later many canals, creating a regular network, were dug to the Atlantic and Gulf in order to further control the water from the lake and to prevent flooding.

At present, Okeechobee's function is to balance the water that it receives from natural runoff from the north and back pumping from the south, to distribute portions of it through canals when and where it is needed, and to retain other portions for environmental needs and irrigation. Because of its great area, the lake holds a very large amount of water in spite of its shallowness. The lake level should be raised to 19.5 feet above mean sea level in order to meet the required needs for the next 30 years. The lake also serves as a receptacle for the back-pumped drainage of surplus storm water from adjoining farm areas. The canal system serves not only for drainage purposes, but also for the irrigation of agricultural lands.

Another important factor in the elaborate water management plan of the region is the storage capacities of Conservation Areas 1, 2, and 3. These are areas that have been surrounded by levees in order to store surplus water and prevent overdrainage and waste. Figure 73 shows the locations of Lake Okeechobee, the conservation areas, and various levees and canals. It also shows the pump stations, which are capable of pumping 5,000 cubic feet per second (2,250,000 gal-

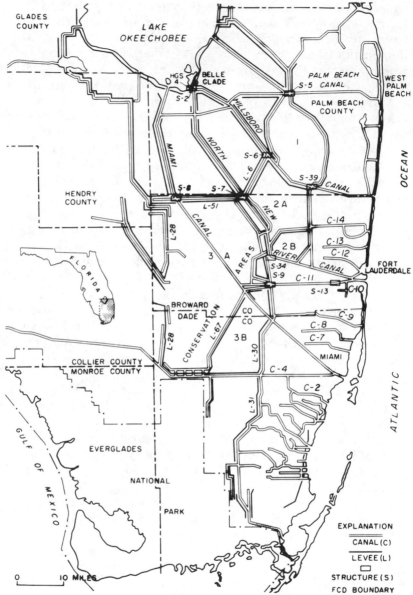

Fig. 73. *Map of southeastern Florida showing canal and levee system of the Central and Southern Florida Flood Control District.*

lons per minute). These pumps can move water southward to the conservation areas or northward to Lake Okeechobee. During the dry season, water stored in the conservation areas can be released to the coastal regions in order to offset water losses and maintain proper water levels above salinity controls. An example of the effective work of the system occurred in 1968. During May and June of that year, a record rainfall occurred, and the area was in danger of extensive flooding. Rapid routing of excess rainfall to the sea averted the potential damage to urban areas. In 1971, a year of extreme drought, water was routed from Lake Okeechobee and the conservation areas to the Everglades National Park and urban areas to reduce the drought effects.

Saltwater Encroachment

Wherever aquifers come in contact with bodies of salt water, there is likely to be a struggle between the freshwater from the land and the salt water of the sea. During dry periods, salt water tends to enter the porous areas of the rock below sea level, displace the freshwater, and thus encroach on the land. During wet periods, freshwater tends to push seaward and displace and override the encroaching seawater. The amount of freshwater available and the level of the water table at the coast are critical factors in this struggle.

In the late 1890s and early 1900s, conditions along the southeastern shoreline of Florida were much different than at present. Early reports indicate that during the rainy season the Everglades were flooded as far east as the western flanks of the coastal ridge, and water gushed through sloughs across the ridge. In other words, the only section that was above water here was the high area of Coconut Grove in Miami. It is evident that freshwater was holding its own in the battle with saltwater encroachment in those days.

There are many reports that during that period fresh groundwater flowed eastward through the rocks to the shoreline and appeared as springs gushing up to the surface of Biscayne Bay. Also, numerous wells drilled near the shore produced freshwater. Wells 50 to 60 feet in depth located at the mouth of the Miami River produced freshwater in abundance. In other words there was very little, if any, encroachment by salt water from Biscayne Bay into the porous rocks along the shore because the water table was high.

Then changes began to take place. With the increase in population and need for the reclamation of swampy and periodically flooded lands, a statewide program of dredging and canal development began in 1907. In the Miami area the Miami River was dredged and the Miami Canal was extended to a point about four miles farther west of the head of the river. This made it possible for the surface water that was stored behind the coastal ridge, as well as the groundwater that was stored in the underlying rocks, to spill out into Biscayne Bay as waste. When we talk of the Everglades today, we mean an uninhabited and swampy region many miles west of the coastal ridge, but the Everglades originally extended eastward right up to the base of the ridge. It is hard to realize that the well-developed and well-populated lands, including much of present-day Coral Gables and Miami International Airport, were once a part of the Everglades. The dredging of a few canals created this change in a very short period of time.

The canals have served to expose a large area of good, dry land to development. Whenever the natural balance is changed by man, it often has both favorable and unfavorable aspects. The drainage of surface water from the Everglades and groundwater from the ridge caused a permanent drop in the water table, and this was the beginning of new and at times very serious problems. Wells near the shoreline that had previously produced freshwater began to be contaminated with salt water. Some of these wells had to be abandoned and new ones had to be drilled farther from the coast.

As soon as the canals were dredged, salt water rushed inland and sea encroachment began in earnest. This took place not only at the surface, but also in the underlying rocks of the Biscayne aquifer. The maps of figure 74 show the progressive saltwater encroachment in the Miami area between 1904 and 1953. In order to help stem encroachment, dams were placed in the canals in strategic places to reduce wastage and to maintain sufficiently high water levels to prevent further intrusion.

Controls established during the last 25 years have been

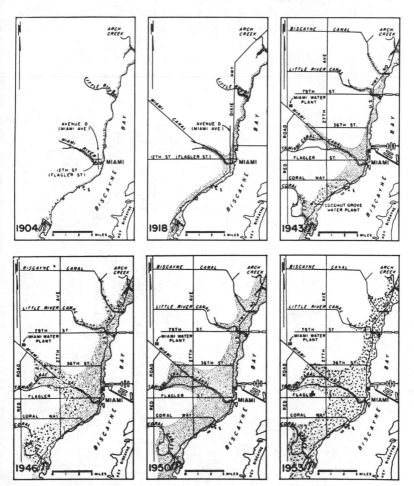

Fig. 74. *Progressive saltwater encroachment in the Miami area from 1904 through 1953. Stippling shows extent of areas that have chloride concentration approximating 1,000 ppm or more at the base of the aquifer.*

fairly effective, and the rapid encroachment of salt water previous to that time has been stemmed fairly well. Figure 75 shows the inland extent of saltwater intrusion in 1967 and indicates that a relatively stable position has been created. This situation could be changed at any time depending on a number of factors, especially by the amount of rainfall. The battle between fresh and salt water is and will be a continuing one.

Finally, it should be emphasized that although the water needs of southeastern Florida have to the present time been

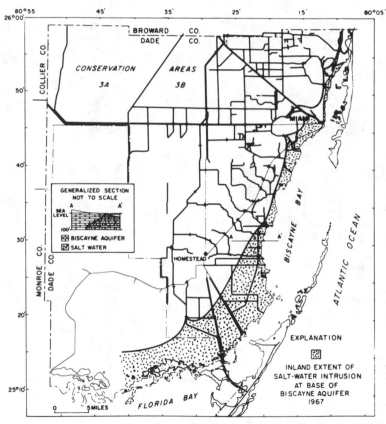

Fig. 75. *Map showing the inland extent of saltwater intrusion in Dade County in 1967.*

fairly adequately met, there is no doubt that improvements of the present system must be made very soon in order to supply the requirements of the near future. The U.S. Corps of Engineers estimated that by 1976 the demands of the lower east coast would exceed the available supply. An intensive cooperative study to develop the most favorable and effective plan to solve what might become a water crisis is now in progress. Among the various plans that will have to be considered is the reduction of runoff to the ocean and the back pumping of excess storm drainage into an enlarged Lake Okeechobee and the conservation areas. In addition, serious consideration must also be given to the retention of properly treated sewage in the canal-aquifer-conservation system.

11| *A Few Things to Remember*

In any study of the chief geologic features of the southern tip of Florida, there are some main points that should be stressed. By far the most important type of rock exposed in this area is limestone. We have seen that limestones are made in several different ways. Some are made organically by both animals and plants and others inorganically (or purely chemically) by the precipitation of calcium carbonate from solution in seawater. The best examples of the former found here are the coral reef limestones of the Florida Keys and the bryozoan limestones that occupy so much of the area of the Everglades. The former are composed of the hard parts of corals and fragments of the shells of numerous other animals and plants. The inorganically produced limestones are represented chiefly by the oolites of the Atlantic coastal ridge and the lower Florida Keys. Not only can we see these rocks, but we can see where and how similar rocks are being made today. The Florida reef tract with its various species of corals and associated animals and plants provides us with a front-row view of this production workshop. The unconsolidated oolite mound just south of Bimini on the Great Bahama Bank gives us a clear picture of the way in which our Atlantic coastal ridge was formed.

This region of southern Florida with its surrounding marine waters is a vast limestone factory, with much more limestone forming here than in any other part of our country. We might even think of it as a place where there are several factories, each one producing a different kind of limestone. In addition to those already mentioned, we can add the huge factory of Florida Bay where fine-grained lime muds are accumulating. Each kind of limestone produced is dependent upon the environmental conditions (both physical and biological) under which it exists.

Study of the oceans and the sea floor provides us with a tremendous amount of information concerning how the land above the water was made—the present being a key to the past. This is especially important when we realize that probably more than 75 percent of all the surface rocks of the continents were made beneath the sea.

Certainly one of the major concepts for an understanding of this region is the tremendous part played by the worldwide rising and lowering of the ocean level during the Pleistocene or Ice Age. When the level was low, as it was during the time of the Wisconsin glacial period, erosion of the newly exposed ocean bottom resulted in great changes in its topography. Much of the surface was washed away. When the sea level began its slow rise, new deposits were made and the ocean bottom was covered by what are today relatively recent rock materials.

We can therefore recognize the existence of rock formations of two different periods in our region—those that were made before the last period (Wisconsin) of glaciation and those made since then. Those that belong to the former group are the Miami limestone and the Key Largo limestone. These were made during the warm climatic period (Sangamon) when the level of the sea stood about 25 feet higher (about 100,000 years ago) than at present. With the ushering in of the cold Wisconsin period the sea level fell and erosion began to eat into the newly exposed sea bottom rocks. With the return of warmer conditions the sea level rose and new deposits were made on the eroded surface. The chief examples of these are (1) the living coral reefs and associated lime sands and muds of the coral reef tract, (2) the longshore sands that provide the foundation of the sedimentary barrier islands of the east coast, (3) the limey (Perrine) marl found in the "glades" of the Atlantic coastal ridge, (4) the lime muds and sands of Florida Bay, and (5) the longshore sands and oyster bars that form the foundation of the Ten Thousand Islands. All of these are very young, many of them having been deposited during the last 5,000 years.

With a knowledge of just these few things, a person will have an intelligent understanding of the making of South Florida. This is truly a "Land from the Sea."

Selected Bibliography

Ball, M. M. "Carbonate Sand Bodies of Florida and the Bahamas."
    Journal of Sedimentary Petrology 37 (No. 2, 1967):556-591.
Ball, M. M.; Shinn, E. A.; and Stockman, K. W. "The Geologic Effects
    of Hurricane Donna in South Florida." *Journal of Geology* 75
    (No. 5, 1967):583-597.
Bayer, F. M., and Work, R. C. "Littoral Marine Life of South Florida."
    Guide Book for Field Trip No. 7, for Geological Society of
    America annual convention in Miami, 1964.
Cooke, C. W. "Geology of Florida." Tallahassee, Fla., State Geological
    Survey Bulletin 29, 1945.
Cooke, C. W. "Scenery of Florida." Tallahassee, Fla., State Geological
    Survey Bulletin 17, 1939.
Davis, John H., Jr. "The Ecology and Geologic Role of Mangroves in
    Florida." Washington, D.C., Carnegie Institution, Publication No.
    517, 1940.
Ginsburg, R. N. "South Florida Carbonate Sediments." Guide Book for
    Field Trip No. 1, for Geological Society of America annual con-
    vention in Miami, 1964. Reprinted as *Sedimenta* 2. Miami: Com-
    parative Sedimentology Laboratory, University of Miami, 1972.
Hoffmeister, J. E., and Multer, H. G. "Fossil Mangrove Reef of Key
    Biscayne, Florida." *Geological Society of America Bulletin* 76
    (1965):845-852.
Hoffmeister, J. E., and Multer, H. G. "Geology and Origin of the Flor-
    ida Keys." *Geological Society of America Bulletin* 79
    (1968):1487-1502.
Hoffmeister, J. E., and Multer, H. G. "Growth Rate Estimates of a
    Pleistocene Coral Reef of Florida." *Geological Society of Amer-
    ica Bulletin* 75 (1964):353-358.
Hoffmeister, J. E.; Jones, J. I.; Moore, D. R.; and Multer, H. G. "Living
    and Fossil Reef Types of Southern Florida." Guide Book for
    Field Trip No. 3, for Geological Society of America annual con-
    vention in Miami, 1964.

Hoffmeister, J. E.; Stockman, K. W.; and Multer, H. G. "Miami Limestone of Florida and its Recent Bahamian Counterpart." *Geological Society of America Bulletin* 78 (1967):175-190.

Meyer, F. W. "Preliminary Evaluation of the Hydrologic Effects of Implementing Water and Sewerage Plans, Dade County, Florida." Tallahassee, Fla., U.S. Geological Survey in cooperation with the Water Control Division, Dade County Department of Public Works, Open File Report 71003, 1971.

Multer, H. G., and Hoffmeister, J. E. "Subaerial Laminated Crusts of the Florida Keys." *Geological Society of America Bulletin* 79 (1968): 183-192.

Newell, N. D.; Imbrie, J.; Purdy, E. G.; and Thurber, D. L. "Organism Communities and Bottom Facies, Great Bahama Bank." *American Museum of Natural History Bulletin* 117 (1959):185-228.

Parker, G. G.; Ferguson, G. E.; Love, S. K.; and others. "Water Resources of Southern Florida with Special Reference to Geology and Ground Water of the Miami Area." Washington, D.C., U.S. Geological Survey, Water Supply Paper 1255, 1955.

Price, W. Armstrong. "Development of the Basin-in-Basin Honeycomb of Florida Bay and the Northeastern Cuban Lagoon." *Gulf Coast Association of Geological Societies* 17 (1967):368-399.

Puri, Harbans S., and Vernon, Robert O. "Summary of the Geology of Florida and a Guide Book to the Classic Exposures." Tallahassee, Fla., State Geological Survey, Special Publication No. 5, 1964.

Scholl, David W., and Stuiver, Minze. "Recent Submergence of Southern Florida." *Geological Society of America Bulletin* 78 (April, 1967):437-454.

Schroeder, M. C.; Klein, H.; and Hoy, N. D. "Biscayne Aquifer of Dade and Broward Counties, Florida." Tallahassee, Fla., State Geological Survey, Report of Investigations No. 17, 1958.

Shinn, E. A. "Coral Growth-Rate, an Environmental Indicator." *Journal of Paleontology* 40 (No. 2, 1966):233-240.

Smith, F. G. Walton. *Atlantic Reef Corals.* Coral Gables, Fla.: Univ. of Miami Press, 1971.

Stanley, S. M. "Paleoecology and Diagenesis of Key Largo Limestone, Florida." *American Association of Petroleum Geologists Bulletin* 50 (No. 9, 1927):

Stockman, K. W.; Ginsburg, R. N.; and Shinn, E. A. "The Production of Lime Mud by Algae in South Florida." *Journal of Sedimentary Petrology* 37 (No. 2, 1967):633-648.

Vaughan, T. Wayland. "A Contribution to the Geologic History of the Floridian Plateau." Washington, D.C., Carnegie Institution, Papers from the Tortugas Laboratory, Vol. 4, 1910, Publication No. 133.

Wanless, Harold R. "Sediments of Biscayne Bay—Distribution and Depositional History." Miami, Fla., Univ. of Miami Institute of Marine Science, Technical Report, August, 1969.

Yonge, C. M. Y. *A Year on the Great Barrier Reef.* New York: G. P. Putnam's Sons, 1930.

Index

Acropora cervicornis, 65, 75, 76, 80, 81, 88, 93
Acropora palmata, 75, 76, 78, 89, 93, 96
Anastasia formation, 22, 23, 125
Angelfish Creek, 99, 101
artesian aquifers, 127
Atlantic coastal ridge, 27-29, 32, 39, 41, 45-47, 92, 101, 136
Atlantic Oceanographic Laboratory (NOAA), 52
Avenidia nitida (black mangrove), 57-61

Barnes Sound, 97, 99, 101
Bayshore Drive, 29, 30
Bear Cut, 53, 54, 99
bedrock, 29, 34, 52, 112, 116, 122
Big Cypress Swamp, 27
Big Pine Key, 83, 84, 85, 86, 87, 91, 94
Bimini, 41, 42, 47
Biscayne aquifer, 123-126, 132
Biscayne Bay, 50, 51, 63, 97-102, 105, 126, 131
Broward County, 24, 27, 33, 40, 122, 124
bryozoans, 35-38, 39-46

Caesar's Creek, 99, 101
Caloosahatchee marl, 22, 23
canal and levee system, 128-131

Cape Florida State Park, 53
Cape Romano, 116, 118
Card Sound, 97, 99, 101
Carysfort Lighthouse, 75, 76
Coastal Lowlands, 21
Coconut Grove, 27, 29, 30, 131
Collier County, 24, 40
Coral Gables, 132
Coral Gables Canal, 29
coral environments, 66-68, 75, 77, 78
coral growth rate, 79-81, 89-90
coral reefs: 63-69;
  atolls, 68, 69;
  back reef zone, 69, 72, 77, 78, 92, 94;
  barrier, 68, 69;
  distribution, 68;
  fringing, 68, 69;
  outer, 69, 72, 73, 75, 89, 92, 94, 95
  patch, 69, 72, 73, 77, 89, 92, 93, 94, 95
coral rock, 29, 86
corals: 63-69; 73-75;
  black coral, 65;
  brain coral, 66, 75, 76, 78, 88;
  moosehorn coral, 75, 78, 93, 96;
  red coral, 64-65;
  rose coral, 78;
  staghorn coral, 76, 78, 80, 81, 93;

star coral, 74, 76, 78, 79-81, 89-90, 93;
stony coral, 65
core drilling, 38-39, 87
cross-bedding, 30, 41, 42, 45
Cross Key Canal, 89, 90, 91

Dade County, 24, 33, 40, 53, 122, 124, 126
Davis, John H., Jr., 121
Diploria, 66, 76, 88
Dry Tortugas, 86

Elliott Key, 99, 101, 103
Everglades, 27, 28, 34, 35, 39, 40, 41, 42, 45, 46, 47, 113, 129, 136
Everglades National Park, 128, 131

Featherbed Bank, 99
Flamingo, 40
flood control districts, 128
Florida Bay, 37, 73, 83, 84, 97, 106-115
Florida Keys, 25, 39, 45, 63, 70, 83-96, 136
Florida reef tract, 63, 69, 70-82, 95
Floridan Plateau, 19, 20
Fort Lauderdale, 27, 40, 125
Forty Mile Bend, 40
Fort Thompson formation, 23, 24, 125, 126
French Reef, 71, 73

geologic map, 22
Geologic Time Table, 23
Ginsburg, R. N., 110
"glades" (transverse valleys): 46, 47, 137
Grassy Key, 87
Great Bahama Bank, 41, 42, 43, 44, 47, 49, 91, 92, 136
Great Barrier Reef, 65, 86
Grossman Hammock State Park, 127

groundwater, 123, 124, 126, 127, 129

Homestead, 27, 125
Hurricane Betsy, 62

John Pennekamp Coral Reef State Park, 71, 73, 82

Key Biscayne, 50, 51, 52, 53, 55, 86, 99;
petrified forest of, 62
Key Largo, 73, 83, 84, 87, 103
Key Largo Dry Rocks, 71, 73, 76
Key Largo limestone, 22, 23, 39, 85-96, 100-101, 125
Key Largo limestone ridge, 23, 24, 100, 101, 103, 104, 105
Key Vaca, 84
Key West, 73, 84, 87
Kissimmee River, 128

Lake Okeechobee, 22, 124, 128-129, 131, 135
Lignum Vitae Key, 83
limestone formation, 108, 136-137
Long Key, 84
Looe Key, 95
Lostman's River, 40
Lower Keys, 85, 86, 91-92

mangroves, 57-61, 83, 105, 106, 113, 115, 118-121
mangrove rock reef, 55, 56, 62
Marion Park, 91
Marquesas Keys, 86
Matecumbe Keys, 73, 83-84
Miami, 27, 32, 41, 125, 131, 133
Miami Beach, 50, 51, 52, 86, 99, 101, 104
Miami limestone, 22, 23, 24, 35, 39, 41, 44, 47, 85, 91, 92, 125;
oolitic facies, 39, 40, 41, 86, 92, 100, 101, 104, 106, 113;
bryozoan facies, 39, 40, 41, 116

Miami River, 132
Miami Seaquarium, 52
Monroe County, 28, 124
Monastrea annularis, 76, 77, 78, 79, 81, 88, 89, 90, 93
Mosquito Bank, 71, 73, 77

National Marine Fisheries Service (NOAA), 52
National Oceanic and Atmospheric Administration, 52
Newfound Harbor Key, 86

Ocean Key, 47, 48
Old Rhodes Key, 99, 101
ooids, 31, 32, 38, 41, 47, 48, 49
oolite, 30, 31, 32, 33, 34, 39
oolite mining, 33, 47, 48
oolite sand (unstable), 41, 43, 46, 91
oyster bars, 118, 119, 120, 138

Palm Beach County, 24, 122, 124
Pamlico sand, 23, 25, 26
Pamlico shoreline, 25, 45
pellets, 38
Penicillus, 109-112, 115
Perrine marl, 46, 137
Plantation Key, 83
Porites asteroides, 77, 78, 88
Porites divaricata, 78, 79
Porites porites, 88

Rhizophora mangle, (red mangrove), 57, 120
Rickenbacker Causeway, 53, 54
Rosenstiel School of Marine and Atmospheric Science, 52

Safety Valve, 97, 99

saltwater encroachment, 131, 133, 134
Sandy Cay, 43
Sangamon interglacial, 94, 104, 137
Sarasota, 24
Schizoporella floridana, 35, 36, 37, 38, 42, 44
sedimentary barrier islands, 50-62
Seven Acres, 91
Shinn, E. A., 80, 110
Scholl, D. W., 121
Silver Bluff, 29, 30, 104
Soldier Key, 52, 63, 83, 85, 86, 101
Spanish Harbor Key, 84, 91
spoil banks, 34, 38
Stock Island, 91
Stockman, K., 110
Straits of Florida, 63, 70
Stuiver, M., 121

Tamiami formation, 22, 23, 24, 116, 125, 127
Tamiami Trail, 40
Ten Thousand Islands, 24, 116-120, 138

University of Miami, 52
Upper Keys, 85, 86-91, 92

Virginia Key, 50, 51, 52, 53, 86, 99
Vizcaya, 32, 33

water table, 122-123
Windley Key, 83, 87
Wisconsin glacial, 25, 94, 104, 137
worm tubes, 38, 56-57

Yonge, C. M. Y., 65